BIOLOGY AND ECOLOGY OF THE VENOMOUS CATFISHES

Biology and Ecology of Marine Life Series

BIOLOGY AND ECOLOGY OF THE VENOMOUS CATFISHES

Ramasamy Santhanam, PhD

APPLE ACADEMIC PRESS

First edition published 2024

Apple Academic Press Inc.
1265 Goldenrod Circle, NE,
Palm Bay, FL 32905 USA

4164 Lakeshore Road, Burlington,
ON, L7L 1A4 Canada

CRC Press
2385 NW Executive Center Drive,
Suite 320, Boca Raton FL 33431

4 Park Square, Milton Park,
Abingdon, Oxon, OX14 4RN UK

Apple Academic Press exclusively co-publishes with CRC Press, an imprint of Taylor & Francis Group, LLC

Library and Archives Canada Cataloguing in Publication

Title: Biology and ecology of the venomous catfishes / Ramasamy Santhanam, PhD.
Names: Santhanam, Ramasamy, 1946- author.
Series: Santhanam, Ramasamy, 1946- Biology and ecology of marine life book series.
Description: First edition. | Series statement: Biology and ecology of marine life series | Includes bibliographical references and index.
Identifiers: Canadiana (print) 20230556078 | Canadiana (ebook) 20230556132 | ISBN 9781774914526 (hardcover) | ISBN 9781774914533 (softcover) | ISBN 9781003409298 (ebook)
Subjects: LCSH: Catfishes. | LCSH: Poisonous fishes—Venom.
Classification: LCC QL637.9.S5 S26 2024 | DDC 597/.49—dc23

Library of Congress Cataloging-in-Publication Data

CIP data on file with US Library of Congress

ISBN: 978-1-77491-452-6 (hbk)
ISBN: 978-1-77491-453-3 (pbk)
ISBN: 978-1-00340-929-8 (ebk)

BIOLOGY AND ECOLOGY OF MARINE LIFE BOOK SERIES

SERIES AUTHOR

Ramasamy Santhanam, PhD
Former Dean, Fisheries College and Research Institute,
Tamil Nadu Veterinary and Animal Sciences University,
Thoothukkudi, India
Email: rsanthaanamin@gmail.com

Books in the Series:

- Biology and Culture of Portunid Crabs of the World Seas
- Biology and Ecology of Edible Marine Bivalve Molluscs
- Biology and Ecology of Edible Marine Gastropod Molluscs
- Biology and Ecology of Venomous Marine Snails
- Biology and Ecology of Venomous Stingrays
- Biology and Ecology of Toxic Pufferfish
- Biology and Ecology of Bioluminescent Marine Fishes
- Biology and Ecology of the Venomous Catfishes

ABOUT THE AUTHOR

Ramasamy Santhanam, PhD
Ex-Dean, Fisheries College and Research Institute,
Tamil Nadu Veterinary and Animal Sciences University,
Thoothukudi, India

Ramasamy Santhanam, PhD, is the Ex-Dean of the Fisheries College and Research Institute (FC & RI), Tamil Nadu, Dr. J. Jayalalithaa Fisheries University, India. He has about 50 years of teaching and research experience in marine/fisheries sciences. He is currently serving as a fisheries expert for various government and non-government organizations in India and abroad. Dr. Santhanam has so far published 32 books on various aspects of marine life, fisheries environment, fisheries biology, and aquaculture. He was a member of the American Fisheries Society, United States; World Aquaculture Society, United States; Global Fisheries Ecosystem Management Network (GFEMN), United States; and the IUCN's Commission on Ecosystem Management, Switzerland.

CONTENTS

ABBREVIATIONS

CK	creatine kinase
CK-MB	creatine kinase-myocardial brand
CPR	cardiopulmonary resuscitation
GSI	gonadosomatic index
LD	lethal dose
TCA	trichloroacetic acid

PREFACE

Catfish of the order Siluriformes are one of the largest orders of teleosts, containing about 4100 species and represent 12% of all teleost fishes and 6.3% of all vertebrates. This important group of fishes is used as food, for décor and for sport fishing. Catfish are found distributed worldwide. However, they are most abundantly distributed in the tropics of Asia, Africa, and South America. While freshwater ecosystems contain the maximum number of species, only two families (Ariidae and Plotosidae) of catfish are common in coastal waters due to their diversity and worldwide distribution; these catfish serve as models to evolutionary biologists and ecologists besides serving as valuable species for biogeographical studies.

Like jellyfish, scorpions, and snakes, several species of catfish have been reported to be venomous with their painful stings. The major groups of persons at risk for catfish envenomation are fishermen and water sports participants. It is estimated that 1625 species of catfish are venomous although venomology has been studied for only 158 species. While some species of catfish cause mild envenomation, several species have been reported for severe envenomation associated with secondary bacterial infections. The symptoms related to catfish envenomation include local intense pain, edema, erythema, paleness, and cutaneous necrosis. Notable species of catfish causing serious envenomations include the Arabian Gulf catfish (*Arius thalassinus),* Carolina madtom (*Noturusfuriosus*), and Oriental catfish (*Plotosuslineatus*).

Though several books relating to venomous fauna are available in the market, a comprehensive volume on the Biology and Ecology of Venomous Catfishes is yet to be published until now. This title, the first of its kind, is the need of the hour. It deals with the biology and ecology of venomous and traumatogenic freshwater and marine catfish diversity; aspects of venomology of catfish such as venom apparatus and venom characteristics; catfish envenomation and its treatment and management; and pharmaceutical value of catfish venoms. It is strongly believed that this volume will serve as a standard text-cum-reference for students and researchers of several disciplines, such as marine biology, fisheries science, and aquatic biology; as a guide for physicians and environmentalists; and as a valuable reference book for the libraries of colleges and universities of the world.

The comments and suggestions of Dr. K. Venkataramanujam, ex-dean of the Fisheries College and Research Institute (Tamil Nadu Veterinary and Animal Sciences University, Thoothukudi, India) on the manuscript are greatly appreciated. I am also grateful to all my international friends for sharing their catfish images for this title. The secretarial assistance provided by Mrs. Albin Panimalar Ramesh is gratefully acknowledged.

CHAPTER 1

INTRODUCTION

ABSTRACT

In this chapter, the possible number of venomous species of fishes and tetrapods occurring globally; the venom types and clinical manifestations of stings of different fish groups; the ecology and biodiversity of venomous catfishes; and envenomation of catfishes of both freshwater and marine environments are dealt with.

Venomous animals of aquatic environments often pose problems for commercial fishermen, anglers, and seafood processors. Among these animals, many species of fishes are known for their envenomation and toxicity. The common venomous fish species known for envenomation are catfish (*Plotosus lineatus*), stonefish (*Synanceia verrucosa*), lionfish (*Pterois volitans*), stingray (*Taeniura meyeni*), waspfish (*Hypodytes rubripinnis*), rabbitfish (*Siganus canaliculatus*), and silver scat (*Selenotocota multifasciata*). Among these fishes, catfish, stonefish, and lionfish have been designated as first, second, and third, respectively, as per their envenomation (Chan et al., 2010). Among the various groups of fishes, the ray-finned fishes (Actinopterygii) rank the first in venomosity followed by catfishes. The diversity of venomous species of fishes and tetrapods; and the venom type and symptoms of various venomous fish groups are given below.

Biology and Ecology of the Venomous Catfishes. Ramasamy Santhanam, PhD (Author)

Globally Believed-to-be-Venomous Species of Fishes and Tetrapods

Fishes and tetrapods	Maximum species
Cartilaginous fishes	200
Spiny-rayed fishes	650
Catfishes	1625
Ray-finned fishes	2275
Lobe-finned fishes and tetrapods	685

Source: Wright (2009).

Venom Type and Clinical Manifestations of Stings of Different Fish Groups

Fish group	Venom type	Symptoms
Catfish	Crinotoxins	Hemolytic, vasoconstrictive necrotic, inflammatory, and severe pain
Stonefish	Verrucotoxin and stonustoxin	Severe pain, swelling, numbness, hypotension, cardiac arrhythmia, and vomiting
Waspfish	Proteins	Swelling, erythema, and extreme pain
Lionfish	Prostaglandins and thromboxane	Swelling and burning pain
Stingray	Serotonin phosphodiesterase, and nucleotidase	Wounds, lacerations, pain, and discoloration near the wound

Source: Chan et al. (2010).

1.1 IMPORTANCE OF CATFISHES

Catfish (order Siluriformes) comprises 34 extant families with 400 genera and about 3000 species. While a majority of these species live in freshwater habitats and a few species are found distributed in the tropical, subtropical, and temperate seas. Generally, these catfishes are without scales although heavy plates of tough, armored skin are seen in some species. Two to four pairs of whiskers or barbels are seen around their mouth. In several species, dorsal fins and gills may have venomous or traumatic spines. A sizable number of catfish species has been reported to be out of water for considerable time as long as their skin is moist by their external layer of mucus (https://science.jrank.org/pages/1268/Catfish.html). Size of these fishes may vary from *Corydoras hastatus* (pygmy Corydoras) which is about 2 cm long, to *Pangasianodon gigas* (giant catfish) the size of which is about 20 m. Catfish are of great use in recreational fisheries and aquaculture. The well-known channel catfish (*Ictalurus punctatus*) is produced in larger quantities through aquaculture especially in the United States of America. Several air-breathing catfish (Clariidae) and shark catfish (Pangasiidae) species assume greater importance in aquaculture, especially in Africa and Asia. Catfish, such as *Corydoras sterbai* and *Corydoras pygmaeus*; and armored suckermouth catfish (*Hypostomus plecostomus*) are of great importance in the ornamental fish trade. Many catfish species such as *Clarias batrachus*, and *Clarias gariepinus* serve as important food fish in Asian countries. The Vietnam catfish *Pangasius bocourti*, has attracted the U.S. catfish industry.

1.2 VENOMOUS CATFISHES: SPECIES COMPOSITION AND GLOBAL AND GEOGRAPHICAL DISTRIBUTION

Biodiversity: In regards, the biodiversity of catfish, there are 37 families with currently recognized 3407 species (10.8% of all fishes and 5.5% of all vertebrates).

Species Composition of Venomous Catfishes* of the Order Siluriformes

Family	Recognized species	Venomous species*
Callichthyidae	199	188
Bagridae	210	187
Mochokidae	204	178
Heptapteridae	199	125
Ariidae	147	100
Clariidae	111	97
Siluridae	97	79
Claroteidae	66	70**
Doradidae	88	65
Pimelodida	107	60
Schilbidae	61	55
Pangasiidae	28	28
Plotosidae	40	27
Amblycipitidae	33	27
Pseudopimelodidae	34	26

Table (Continued)

Family	Recognized species	Venomous species*
Anchariidae	6	5
Cranoglanididae	5	3
Chacidae	3	3
Total	1638	1323

*Calculated average values of believed-to-be venomous species of catfish.

**Calculated from assumed total catfish species.

Source: Armbruster (2011); Wright (2009).

Global distribution of venomous catfishes: Catfish are well known to most people of the world as they are distributed globally. While the family Pimelodidae is well represented in South America; Bagridae in Africa; Ictaluridae in the rivers and lakes of North America; Doradidae in the rivers of South America; Plotosidae from the Indo-Pacific region; Ariidae in the tropical to temperate continental shelves around the world; and Siluridae in Europe and Asia (Halstead, 1988). Venomous fishes have been reported to outnumber all other venomous vertebrates (Smith and Wheeler, 2006), and it has also been demonstrated that the diversity of venomous catfish exceeds that of all other venomous vertebrates.

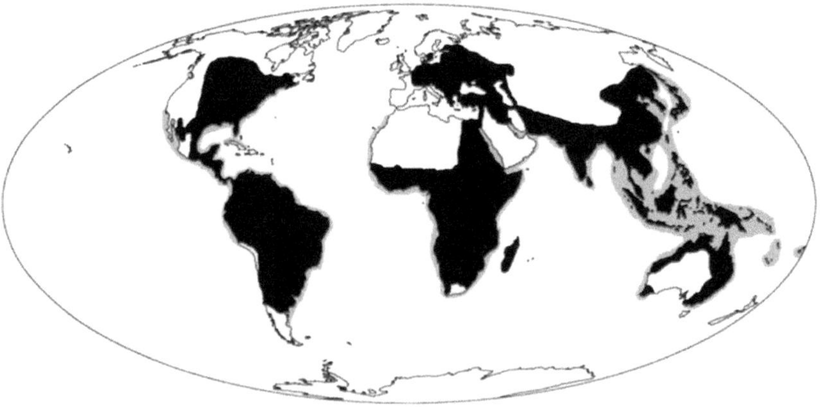

Global distribution of catfishes

(Dark areas are freshwater distributions; gray areas are marine distribution).
Image credit: Jonathan W. Armbruster (Reproduced with permission).

Geographical Distribution of Catfish

Country/Region	Family/Families
Worldwide	Ariidae
North America	Ictaluridae
South America	Aspredinidae, Cetopsidae, and Doradidae, Pseudopimelodidae and Scoloplacidae
Central America: Panama	Astroblepidae, Auchenipteridae, and Callichthyidae, Pimelodidae, Loricariidae, Trichomycteridae, and Heptapteridae,
Mexico	Lacantuniidae
Chile	Nematogenyidae and Diplomystida
Africa	Amphiliidae, Auchenoglanididae, Austroglanididae, Claroteidae, Malapteruridae, and Mochokidae
Madagascar	Anchariidae
Europe and Asia	Siluridae
Asia and Africa	Bagridae, Clariidae, and Schilbidae
South Asia	Amblycipitidae, Akysidae Amblycipitidae, Chacidae, Heteropneustidae, Horabagridae, Pangasiidae and Sisoridae
China and Vietnam	Cranoglanididae
Indo-Pacific, Australia, New Guinea	Plotosidae

Source: Armbruster (2011).

1.3 ECOLOGY OF VENOMOUS CATFISHES

Though the catfish are found in all aquatic biomes worldwide, the species of the families Ariidae and Plotosidae, are unique as they are not only marine, but are also found in brackish and coastal water, and sometimes in exclusively freshwater environment (Haddad, Jr. and Martins, 2006).

1.4 BIODIVERSITY OF VENOMOUS CATFISHES

Though the order Siluriformes contains venomous catfish representatives for nearly three centuries, the earliest report on catfish envenomation was that of the family Ariidae. The venomous catfishes have been grouped with other venomous animals such as bees, spiders, scorpions, snakes, lizards, and jellyfish. It is worth mentioning here that more than half of all venomous

Distribution of Venomous Catfish in Different Aquatic Biotopes

Family	Freshwaters	Freshwaters and marine waters
Akysidae	+	
Amblycipitidae	+	
Anchariidae	+	
Ariidae		+
Bagridae		+
Callichthyidae	+	
Chacidae	+	
Clariidae	+	
Claroteidae	+	
Cranoglanididae	+	
Doradidae		+
Heptapteridae	+	
Ictaluridae	+	
Mochokidae	+	
Pangasiidae	+	
Pimelodidae	+	
Plotosidae		+
Pseudopimelodidae	+	
Schilbidae		+
Siluridae		+

Source: Wright (2009).

fish species are catfishes, which possess venom glands (Wright. http://www.nysm.nysed.gov/research-collections/biology/ichthyology/research/transcriptomic-identification-toxic-components). Among the valid catfish species numbering 3000, 20 families of 1250–1625 species have been reported to be venomous inflicting a painful sting (Armbruster, 2011; Anon., https://www.futurity.org/just-your-garden-variety-poisonous-catfish/). Of the 20 families of catfish, 14 families viz. Akysidae, Heptapteridae, Mochokidae, Anchariidae, Callichthyidae, Pangasiidae, Pimelodidae Chacidae, Claroteidae, Schilbidae, Siluridae, Cranoglanididae, Doradidae, and Pseudopimelodidae) have been shown to contain venomous taxa for the first time; and six families (Amblycipitidae, Ariidae, Bagridae, Clariidae, Ictaluridae, and Plotosidae) have been reported to possess venomous representatives. Further, among these venomous catfish, the stings of the species of Plotosidae (e.g., *Plotosus lineatus*) and Heteropneustidae (*Heteropneustes fossilis*) were reported to be responsible for human fatalities as these fishes possess potent venoms or cause secondary infection of the wound (Halstead, 1988; Wright, 2015). It is also worth-mentioning here that unlike other venomous creatures, only limited studies are available on the venomous fishes in general and catfish in particular (Wright, 2012). Further, although several families of catfish have been suspected to possess venom-producing representatives, only a few families have been confirmed with venomous species.

Venomous apparatus catfishes: Venomous apparatus catfishes: Several species of marine catfish possess three serrated bony stings or spines, that is, one in each of the two pectoral fins and one in the single dorsal fin. These spines, which are lined with venom glands, are found along the front edges of these fins and are normally venomous (Wright. http://www.nysm.nysed.gov/research-collections/biology/ichthyology/research/transcriptomic-identification-toxic-components). Though the spines of these stings are mainly used for defense against predators, they may cause injuries when the fish is provoked. While the majority of the venomous catfish are with stings which possess venomous glandular tissue under the epithelium, some cause traumatic wounds without poisoning as their stings are without poisonous glands next to the sting (Haddad, Jr. and Martins, 2006). Further, the venom glands have been reported to be smaller in bigger fishes. Koerth-Baker (2009) reported that even catfish of more than 27 kg size may not possess venom glands. The secretions of the venomous glands of the catfish may cause severe pain, cardiovascular, neuromuscular, and cytolytic effects in humans. Compared to other venomous organisms, the venoms of fishes in general and catfishes in particular have received comparatively little attention. A thorough knowledge about venom composition, toxin identity, and genetic architecture of venom production of catfishes may be great use, especially to explore the possibilities of employing the venoms of these fishes potentially for the development of marine drugs.

1.5 VENOM APPARATUS AND ENVENOMATION OF CATFISHES

Venomous catfish stings have been considered as a common environmental hazard worldwide. Catfish stings may, therefore, cause both envenomations and severe trauma with lacerations and punctures in humans. These stings may also be responsible for severe pain, cardiac dysrhythmias, arterial hypotension, foreign body retention, and infection in affected patients. Most injuries relating to catfish stings have been reported to the hands of fishermen, and more case reports are available in freshwater than in marine environments. Catfish envenomation-related local manifestations include local intense pain, edema, erythema, and paleness and occasional cutaneous necrosis. Systemic manifestations have been rarely reported during the first symptom hour (acute envenomation phase). While the punctures caused by the stings of catfish are painful, in severe cases, the pain may last for even 6 h particularly when the venom is deposited by the sting. Further, complications occur when the lacerations are associated with intense bleeding or when stings break away from the fish and remain inside the wound. For the risk of secondary infections, catfish from dirty or polluted waters are said to be mainly responsible (Satora et al., 2008). Microbial species such as that of *Streptococcus, Staphylococcus*, *Vibrio, Aeromonas, Pseudomonas*, and *Edwardsiella* are largely responsible for bacterial infections; and for fungal infections, the major causative species is *Sporothrix schenkii* (Haddad, Jr et al., 2008). Another threat relating to catfish exposure is erysipeloid an infectious skin disease (Fish-Handler's disease') which develops following immediate penetration of the bacterial species, *Erysipelothrix rhusiopathiae,* the source of which are mainly fish and pigs. Erysipeloid usually occurs in fishermen and fish traders. This disease has been reported to cause severe pain with oedema and restricted bluish-red skin lesion (Satora et al., 2008). Reports are also available about certain species catfish like *Urinophilus* sp. which are known to cause envenomation by entering into the human urethra. In such cases, surgical removal of such organisms is mandatory. Further, the Egyptian species like *Malapterus* spp. are responsible for fatal electric shock in humans and this is due to the creation of electric current on the surface of their bodies (Blomkalns and Otten, 1999). Routine treatment measures for catfish stings include hot water immersion, prophylactic antibiotics, analgesia etc. (Dorooshi, 2012).

It is worth mentioning here that the majority of aquatic envenomations are mainly due to catfish and this is largely due to their toxic epidermal secretions or venomous spines. It is also worth mentioning here that >90% of envenomations are punctures. Further, it has been reported that 12% of catfish envenomations is caused in people during their swimming or walking on the beach; and 88% is invariably associated with humans involved in fishing activities. Though a variety of commercial catfish-handling devices are available, all of them are of questionable assistance. It is also advisable to avoid barefoot wading, particularly in slow-moving waters. Catfish-associated superficial scratches and lacerations can be washed with antibacterial soap, dressed with antibiotic ointment, and observed. Wounds that are more serious, however, should be treated in hospitals. Before all these, one should have a fair knowledge about the catfish, especially about their habitats, biology, envenomations, and treatment.

KEYWORDS

- **venomous fishes**
- **venomous tetrapods**
- **catfish biodiversity**
- **catfish distribution**
- **envenomation**

CHAPTER 2

BIOLOGY OF CATFISHES

ABSTRACT

The various aspects of biology of catfish are given in this chapter. Aspects such as morphological characteristics, food and feeding and respiration of catfishes of different catfish families; adaptations possessed by the catfishes inhabiting freshwater, brackish water, coastal water and deep sea ecosystems; and reproductive biology of freshwater and marine catfish are also dealt with.

Catfishes (Order: Siluriformes) which are about one-tenth of living fishes have been reported to live all over the world except Antarctica. While most catfishes live in freshwater environments, some live in brackish water biotopes such as estuaries, mangrove swamps, and others occur in seas and in deeper ocean water. Catfishes are mostly demersal (bottom-dwelling). They are solitary and are nocturnal. On the other hand, some species of this genus inhabit quite away from the bottom and are found active during the day hours. Some species may form schools. Most of the catfishes are bottom feeders feeding mainly on invertebrates; some catfishes eat fish; and others feed on fallen leaves, algae, etc. Some catfish species may eat the blood of other fish. Catfish species usually involve in courtship activity before spawning, or releasing eggs. They are then said to provide some type of parental care, especially by the males. That is males are mostly guarding eggs laid by the females in their mouths until they are hatched.

2.1 CHARACTERISTICS OF CATFISHES

The body of the catfishes may be covered with bony plates or naked. Most of the catfishes possess one to four pairs of threadlike barbels around the

Biology and Ecology of the Venomous Catfishes. Ramasamy Santhanam, PhD (Author)

mouth. When there are four pairs of barbels, they are located as one on maxillary, one on nasal and two on the chin of the fish. Chin or nasal barbels may, however, be absent in some species. Eyes are usually small and barbels largely help in detecting food. Air-breathing organs are present in the catfishes of the families, Clariidae and Heteropneustidae. In most species, caudal fin rays are 17–18. Other characteristic features of these catfishes include the presence of spines in the pectoral fins and at the front of the dorsal fin, or the fin along the midline of the back. The spines of some catfishes are venomous (Anon., https://www.fishbase.de/summary/ordersSummary.php?order=Siluriformes).

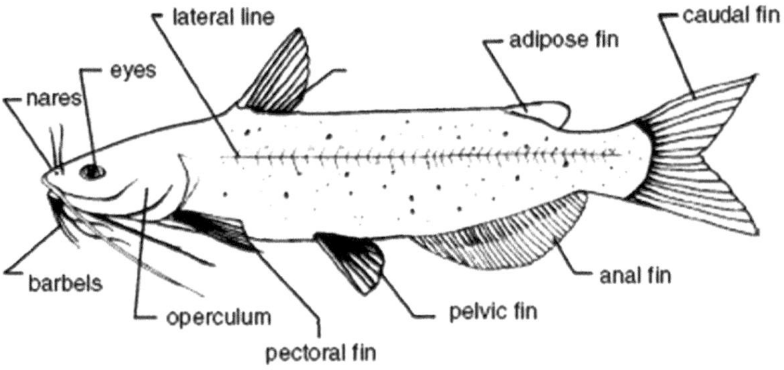

External parts of the Channel Catfish
Image credit: Jimmy L. Avery, PhD. Reprinted with permission.

2.2 CHARACTERISTICS OF CATFISH FAMILIES

Akysidae: The species of this family are cryptically colored, small catfishes (less than 5 cm). They have extremely small eyes and longitudinal row tubercles all over their bodies. The dorsal fin has one leading spine and a short base. Adipose fin is present. There is a strong pectoral spine which is usually serrated. Gill openings are narrow and four pairs of barbels are seen.

Amblycipitidae: They have generally elongated bodies. Dorsal fin is covered by thick skin. In some species, its adipose fin is merging with the tail fin. Barbels are in four pairs. The lateral line is incomplete.

Amphiliidae: They have an elongate appearance with bony plates. Dorsal and anal fin bases are short. There is no spine in dorsal fin. Most species lack pectoral spine. Occasionally, a short spine is seen in adipose fin. Most species attain a length of less than 12 cm.

Anchariidae: The fishes of this family are known for their much reduced anterior nuchal plate and fringed barbels.

Ariidae: These fishes are known for their mouth-brooding. Males carry large eggs in their mouth until they are hatched. Caudal finis forked and an adipose fin is present. There are two to three pairs of barbels. Nasal barbels are invariably absent. Head and dorsal fin are with bony plates. Both pectoral and dorsal fins bear a protruding spine. A leading spine is seen in both pectoral and dorsal fins.

Aspredinidae: The skin of aspredinids is keratinized and is shed like in reptiles. Anterior part of body is depressed. Body of the fish is generally naked and large tubercles may be presented in some species. There is no adipose fin. Body is depressed anteriorly and opercular aperture appears only as a slit. In most species, the dorsal spine-locking mechanism is not seen.

Astroblepidae: Body is naked and mouth disc is present. While maxillary and nasal barbels are present in all species, adipose fin may be presented in only some fishes. Dorsal spine is without a locking mechanism. These fishes are capable of climbing waterfalls by using their pelvic fins which are used mainly in a walking motion.

Auchenipteridae: The body of the fishes of this family is naked. There are three pairs of barbels in all species except one, where the maxillary pair is comparatively longer. Dorsal and pectoral fins are with a strong spine. Adipose fin may be presented in only some fishes. These rare, pink fishes come out only at night. Male auchenipterids have a modified anal fin which is used for internal fertilization.

Auchenoglanididae: Fishes of this family are most similar in appearance to mochokids but these fishes differ from the latter in having a longer snout.

Austroglanididae: In these fishes, there are three pairs of barbels (nasal pair, however, is absent). Dorsal and pectoral spines are strong and adipose fin is small.

Bagridae: In these fishes, a spine is seen before dorsal fin. There is a high interspecific variability that is seen in the size of adipose fin. Pectoral spine is serrated. Body is scaleless and barbels are usually in four pairs and are well-developed. Some bagrids have unusual reproductive behaviors.

Callichthyidae: There are two rows of overlapping bony plates on each side of the body of these fishes. The swim bladder is located in a bony case. Mouth is ventral and small. There are one to two pairs of highly developed barbels. Dorsal and pectoral fins are with strong spine. Adipose fin is with spine at anterior border. Fishes of this family are air-breathers and substrate brooders; and after a very active spawning behavior, they attach their eggs to the substrate.

Cetopsidae: These fishes are generally small (< 6–30 cm). Body is naked with three pairs of barbels. Nasal barbels are absent. Adipose fin is absent. The swim bladder is extremely small and is in a bony capsule.

Chacidae: Head of this species is depressed and broad. Posterior body is compressed. Mouth is large and terminal. Eyes are reduced. One short leading spine is seen in dorsal fin. Pectoral fin is with one serrated spine. Adipose fin is continuous with caudal fin. These fishes are commonly called as angler catfishes as they are known to use their maxillary barbels as lures to attract fishes for food.

Clariidae: In these fishes, the dorsal fin is protruding to a greater length of the body. Dorsal fin is without a leading spine. The dorsal fin may be united or discontinuous to the caudal fin. The caudal fin is rounded. Wide gill openings are seen. Barbels are in four pairs. Pectoral and pelvic fins are either small or absent. Some species with their smaller eyes are burrowers. Species of *Clarias* in this family are capable of leaving water to walk to other bodies of water. Unusually, these fishes can breathe air with their gills because they have bony stays within the gills which keep them from collapsing in air.

Claroteidae: The species of *Phyllonemus* of this family are the first known fishes to have bipaternal mouth brooding. Body of these fishes is moderately elongated with four pairs of barbels. Dorsal and pectoral fins are with strong spines. Adipose fin is present.

Cranoglanididae: Fishes of this family have short dorsal fin. Both dorsal and pectoral fins have a single spine. The body is scaleless with four barbels. Its tail fin is much forked. Top of head is with rough bony plates.

Diplomystidae: Fishes of this are the only living catfishes to retain maxillary teeth. Maxilla is well developed with barbels. Dorsal and pectoral fins are with one spine each. Adipose fin is present.

Doradidae: In these fishes, there is a single row of plates with large spikes running down their sides. Swim bladder variation is common in this family and some bladders have many thin diverticula of unknown function. There is a single row of lateral bony plates along the body. Dorsal fins with a spine Barbels are three paired and nasal barbels are absent. Adipose fin is present. Due to their sound production, these fishes get the name 'talking catfishes'.

Heptapteridae: The skin of these fishes is usually naked (scaleless). They have three pairs of barbels and a large adipose fin. Caudal fin is deeply forked.

Heteropneustidae: Bodies of these fishes are elongate and compressed. Head is much depressed with four pairs of barbels. There is an extended air sac extending backwards from gill chamber and this serves as a lung for these fishes. The dorsal fin is short and doesn't have without any protruding spine. Adipose fin is in the form of a low ridge or it may be absent. One of the most powerful toxins is present in the genus *Heteropneustes* of this family. The paired outgrowths of the buccal cavity of these fishes are used to breathe air.

Horabagridae: Fishes of this family largely resemble that of Bgridae and Schilbidae.

Ictaluridae: The ictalurids (particularly Ictalurus and Pylodictis) are probably the most widely studied catfishes in the world. These fishes possess a naked body with eight barbels which are located near their mouths. A venomous spine is seen in both the dorsal and pectoral fins. Palate is toothless. There are four species of blind catfishes in the genera *Trogoglanis* and *Prietella* of this family.

Lacantuniidae: No description defined.

Loricariidae: Fishes of this family possess bony plates on body and they may be rarely absent. Distinct barbels may or may not be seen in their inferior mouth. Papillae are present on the ventral lip. Adipose fin has a spine at its anterior border when it is present. Intestine is relatively long. Males of many species have hypertrophied odontodes over various parts of their bodies, and in some species, females may also have them.

Malapteruridae: Lateral musculature is modified into an electrical organ in these fishes and this organ is strong enough to shock prey; it may also cause quite a bit of pain to humans. Dorsal fin is absent. Adipose fin is located close to caudal fin.

Mochokidae: These fishes often have dramatic patterns of spots which make them attractive aquarium fishes. Adipose fin is usually very large. Both dorsal and pectoral fin spines possess a locking mechanism. In these fishes, there are three pairs of barbels. While the mandibular barbels are with numerous branches, nasal barbels are totally absent. Part of barbels may also be modified into an oral sucker.

Nematogenyidae: Not reported.

Pangasiidae: Fishes of this family are with a compressed body and their small adipose fin is separated from the caudal fin. Dorsal fin is close to head region. Barbels are usually in two pairs. There is a pair of chin barbels and nasal barbels are absent.

Pimelodidae: These are long-whiskered catfishes with naked body. Adipose fin is present. There are three pairs of barbels. Pectoral and dorsal fins may or may not have spines.

Plotosidae: These fishes have eel-like body and pointed or bluntly rounded tail. Barbels are usually in four pairs. Adipose fin is absent.

Pseudopimelodidae: These are small fishes with a wide mouth and short barbels. Eyes are small and are without free orbital margin. A beautiful pattern of coloration with wide dark brown blotches is seen in the trunk portion of fishes of some genera of this family.

Schilbidae: Dorsal fin is totally absent in some species of this family. Adipose fin is seen in most species. Four pairs of barbels are commonly seen in most species. However, anal fins are absent in several species.

Scoloplacidae: These fishes possess two bilateral series of odontode-bearing plates. Dorsal fin is with a stout smooth spine. These fishes have respiratory stomach.

Siluridae: In these fishes, there is no spine before the dorsal fin. Adipose fin is absent. Pelvic fins are small or absent. Anal fin base is very long. There are no nasal barbels. On the other hand, there are one to two pairs of barbels on its lower jaw, and maxillary barbels are usually elongated. Their clear bodies make them difficult to identify in turbid water.

Sisoridae: In some sisorids, an attaching structure made of folds of epidermis is seen on the thorax. There are normally four pairs of barbels. The adipose fin is usually large. In some species, adipose fin is found united to caudal fin. Dorsal fin may or may not have a spine.

Trichomycteridae: Species of this family have naked and elongate body. There are two pairs of barbels on maxilla and barbels are absent on chin. Adipose fin is absent. Opercle has spines. There are no pelvic fins in three lineages viz. Eremophilus, Miuroglanis, and Glanapteryginae. Several genera of fishes of this family are parasitic and are believed to attack the gill tissue of larger fishes. Fishes of the genus *Vandellia* have been reported to enter the urethra of humans thereby causing serious problems in health.

2.3 FRESHWATER CATFISH SPECIES OCCURRING IN MARINE ENVIRONMENTS

Catfishes of brackish water habitats: Although most of the catfishes belong to freshwater systems, a number of catfish families include species that tolerate or thrive in brackish water areas such as estuaries or mangrove swamps. One such typical species is *Siluris glanis* (Siluridae) which is normally found in freshwater rivers but can also be seen in the brackish waters adjoining the Baltic and Caspian seas. To this category, various species of clariid and ictaluriid catfish are also included. These fishes do not possess any specific adaptations to brackish water but they are incredibly hardy and tolerant of sub-optimal conditions. Another important brackish water species is *Ictalurus punctatus* which thrives in low salinity brackish water aquaria. *Mystus gulio*, a large bagrid catfish of South and South East Asia is a typical estuarine catfish species that prefers brackish water habitat. Other species of this category are Siamese tigers (*Datnioides* spp), and banjo catfishes like *Bunocephalus coracoideus, Aspredinichthys, Aspredo,* and *Platystacus* spp. (Anon., https://www.scotcat.com/articles/article84.htm)

Catfishes of coastal and deep seas: Catfishes of salty water include representatives of mainly two families, viz. Ariidae and Plotosidae. While the catfishes of the family Ariidae (about 120 spp.) are mainly of freshwater environments, a number of species of this family are found along the coastlines of all the continents except Europe and Antarctica. On the other hand, Plotosidae is a smaller family with about 30 species and its salty representatives are confined to the Indo-West Pacific (https://www.scotcat.com/articles/article84.htm).

2.4 BIOLOGY AND REPRODUCTIVE BIOLOGY OF TYPICAL CATFISH SPECIES

2.4.1 BIOLOGY OF ICTALURUS PUNCTATUS

Image credit: Cliff, Wikimedia https://creativecommons.org/licenses/by/2.0/deed.en

Common name: Channel catfish

Habitat: This demersal species lives on sand, gravel, and rock bottoms of streams; quiet waters of lakes, reservoirs, and ponds; and medium to large rivers with strong currents. Further, this fish has been reported to enter downstream in the fall season and upstream during spring. It may also be seen in euryhaline waters. The channel catfish is a relatively hardy fish and it can thrive even in adverse conditions (Anon., https://animals.jrank.org/pages/2027/Catfishes-Siluriformes-CHANNEL-CATFISH-Ictalurus-punctatus-SPECIES-ACCOUNTS.html).

Body description: This lean and naked fish has a gently sloping dorsal profile before its dorsal fin. It is grayish blue to nearly black dorsally; and belly is white to silvery. Juveniles are mottled and brownish on their back Albino channel catfish which has peach color is common in certain domesticated strains. Its barbels are arranged in a typical pattern, that is, four barbels under the jaw and one barbel on each end of the maxilla. The nostrils which bear the olfactory organs are present on the top of the head, slightly in front of the eyes. Its olfactory organs are found in its nostrils (nares), which are located on top of the head just in front of the eyes. It has soft-rayed fins; however, pectoral and dorsal fins are with sharp spines. Anal fin is rounded with 24–30 rays. Its characteristic caudal fin is deeply forked although the depth of the fork is highly reduced especially in large breeding males (Anon., http://extension.msstate.edu/content/biology-catfish). It has a maximum length of 1.3 m and a weight of 26 kg and its life span is about 40 years.

Food and feeding: It is a nocturnal feeder and its diet consists of both plant and animal components. In natural waters, it is known to feed near the bottom and on the surface. Its stomach analysis has shown that young catfish feed mainly on aquatic insects and adults on green algae, snails, crawfish, small fish, aquatic plants, and insects. It may also feed on terrestrial insects, even birds. It is worthy of mention here that these fishes detect their food sources through their sense of taste. In these fishes, taste buds are present not only on their external surface but also in the mouth, pharynx, and gill arches. Taste buds are also found on its barbels and gill arches. While these fishes use their taste buds to locate food in turbid waters, they employ eye sight in clear waters.

Respiration: In these fishes, there are four gills on both sides of the head and each gill consists of a double row of slender gill filaments which are arranged in a double row. Each side of the filament a number of thin, small cross plates called lamellae. Both the gills and gill filaments are provided with several blood vessels. While the oxygen absorbed at the lamellar gill surface is carried throughout the body by the blood, the carbon dioxide waste which is also taken by the blood is released into the water column from the lamellar surface.

Growth: Channel catfish are known to grow well in warm waters with temperatures of 75–85°F, albeit they can also survive at temperatures of about 100°F. In these fishes, poor growth has been reported at temperatures less than 65°F and feeding activity is believed to stop at temperatures below 50°F. Adults may tolerate salinities of 0–11 ppt, although the preferred temperatures are less than 4 ppt. Eggs of these fishes hatch at salinities less than 8 ppt (http://extension.msstate.edu/content/biology-catfish). It has been reported that it does not reach 1 pound in size until it is 2–4 years old.

2.4.2 *REPRODUCTIVE BIOLOGY OF CATFISHES*

2.4.2.1 FRESHWATER CATFISH, *ICTALURUS PUNCTATUS*

Spawning and Reproduction: Channel catfish are known to spawn at water temperatures of 75°–85°F. Depending on the location, wild populations of this species may spawn as early as February or in late August. The peak spawning of these fishes has been reported during May, albeit the length and the period of the spawning season may vary according to the weather conditions of their habitats. This species is a cavity spawner and it spawns only in semi-dark localities. Prior to the spawning season, the male catfish builds a nest in holes in the rock, hollow logs or in the nearby river banks by fanning out mud and debris. The built nest is also guarded by the male until the spawning is completed and the fry leave the nest. The female lays its eggs as a gelatinous mass on the bottom of the nest. While the females are believed to spawn only once in a year and are laying about 4000 eggs per pound of body weight, the males spawn more than once. Though channel catfish become sexually mature at 3 years, some individuals have been reported to spawn even at the age of 2 years in aquaria. On the other hand, wild populations usually spawn after 5 years. Eggs hatch in 5–10 days, depending on water temperature. The newly hatched fry possesses a large yolk sac which will take care of the nourishment for about 5 days. After the absorption of the yolk sac, the fry becomes dark colored and are said to be ready to swim and look for their food. It is also reported that the fry gulp air to fill their swim bladders at their first swim-up in order to help them to maintain their buoyancy (Wellborn, https://thefishsite.com/articles/channel-catfish-life-history-and-biology).

2.4.2.2 BRACKISH WATER CATFISH, *MYSTUS GULIO*

Sexual dimorphism: In this species, sexual dimorphism is distinct and prominent. While males has a prominent muscular papilla with a pinkish tip, it is absent in females. Further, females are normally larger than the males (Kumar et al., 2019).

Size and age at maturity: The values of minimum size (shortest length) at maturity of this species have been reported as 5.4 cm in Lake Pulicat;

6.2 cm in Hooghly estuary; 7.9 cm in Matlah and Kulti estuaries; and 8.2 cm in Chilka Lake (Kumar et al., 2019).

Fecundity: The fecundity of female fish of a species is believed to be of great use in the assessment of the abundance and reproductive potential of a fish stock. The values of number of eggs per gram ovary of this species in relation to the size have been reported as follows:

Fecundity (Eggs/g Ovary) of *Mystus gulio*

Size of Fish (cm)	Fecundity (Eggs/g Ovary)
10–12	12,000
11–25	1300–25,000
12–20	400–19,000
12–25	4000–170,000
15–23	6000–141,000
20–22	22,000

Source: Kumar et al. (2019).

Gonadosomatic index: The values of Gonadosomatic index (GSI) which is an indicator of the stage of ovary maturation and degree of ripeness both male and female of this species have been reported to increase from March with a peak in July (Kumar et al., 2019).

Spawning season: The peak spawning season of this species has been reported from several Indian/Bangladeh localities as follows:

Peak Spawning Period of *Mystus gulio* in Indian and Bangladesh localities

Locality/country	Peak spawning period
Indian areas	June to November
Bangladesh	March to November/July
Hooghly estuary (Bay of Bengal)	July
Pulicat Lake (Tamil Nadu state, India)	September to October
Cooum backwater (Chennai, India)	September to October

Source: Kumar et al. (2019).

Breeding periodicity: This species has been reported to spawn only once in a breeding season (Kumar et al., 2019).

2.4.2.3 SEA CATFISH, *ARIOPSIS FELIS*

As in other members of the Ariidae family, males of this species brood the fertilized eggs and larval fish inside their mouths. These males may even shelter small juveniles capable of feeding on their own. It is also interesting to mention here that these males do not feed for a period of 8–11 weeks while carrying these eggs or young fish, the male parent does not feed. However, the young fish feeds normally on the zooplankton which drifts into the male's mouth. This species of catfish attains sexual maturity at the age of two, when females have a length of 4.9–10.4 inches and males have a size of 9.8 inches. Sexually matured females develop flap-like fatty tissues in their pelvic fins. Therefore, the females possess larger sized fins than their counterparts. These enlarged pelvic fins are believed to serve as a site for fertilization of the eggs during spawning and for moving the fertilized eggs to the male for incubation. However, the males may also pick up the fertilized eggs after they settle to the bottom. The spawning season of this species extends from May to August or September in shallow estuarine and brackish waters. The greenish eggs are oval or elliptical with a size of 12–19 mm in diameter. Only 20–65 eggs are normally released per spawning event. The mouth–brooding by the male catfish helps to protect the eggs during incubation and keeps the eggs well oxygenated by the flow of water though the mouth of the males as they breathe. The eggs hatch in a period of about 30 days and the larvae that emerge will have a size range of 29–45 mm in length. These larvae remain in the male's mouth until their yolk sac is absorbed which is within 2–4 weeks. Subsequently, the juvenile fish begins to possess adult characteristics and starts to begin opportunistic feeding and scavenging. However, these juveniles tend to remain with the male parent and they may even take shelter in his mouth to avoid threats for some time. The lifespan of this species may be about 5–8 years (Anon., https://gcrl.usm.edu/public/fish/hardhead.catfish.php).

KEYWORDS

- **biology of catfish**
- **catfish characteristics**
- **catfish family features**
- **reproductive biology**
- **fecundity**

CHAPTER 3

BIOLOGY AND ECOLOGY OF VENOMOUS FRESHWATER CATFISHES

ABSTRACT

The aspects of biology and ecology of venomous freshwater catfishes belonging to a total of 22 families of the Order, Siluriformes viz. Akysidae, Amblycipitidae, Anchariidae, Ariidae, Astroblepidae, Auchenipteridae, Bagridae, Callichthyidae, Chacidae, Clariidae, Claroteidae, Cranoglanididae, Doradidae, Heptapteridae, Heteropneustidae, Ictaluridae, Mochokidae, Pangasiidae, Pimelodidae, Plotosidae, Schilbidae, and Siluridae are dealt with in this chapter.

3.1 ASIAN STREAM CATFISHES (SILURIFORMES: AKYSIDAE)

Acrochordonichthys rugosus (Bleeker, 1846)

Source: Public domain

Common name(s): Asian Banjo Catfish.

Biology and Ecology of the Venomous Catfishes. Ramasamy Santhanam, PhD (Author)

Global distribution: Tropical; Asia: Thailand, Malaysia, and Indonesia.

Ecology: Freshwater; this demersal species inhabits mainly sandy or rocky bottoms in clear, fast-flowing forested streams. It is known to hide occasionally under submerged rocks or logs.

Biology: This species has numerous small nodule-like growths called granulations which are seen along the full length of the body including the head. It has a brown coloration on its head and body. The head is seen interspersed with small dots and blotches of a darker brown and a light brown saddle. Specimens with light brown tan-colored spots and patches are also reported. Head margins are steeply sloping. The snout is convex when viewed dorsally. The adipose fin is with an angular posterior margin. The pectoral spine has serrations on its posterior edge. As a predatory fish, it is found ambushing fishes of the species of *Nemacheilus* and *Glyptothorax*. It is known to reach a maximum length of 11 cm.

Venomous organs: The venom delivery system constitutes the bony spine of this species and is Y shaped, and the venom gland is paired and is in the anterior grooves of the spine (Wright, 2012).

Akysis clinatus (Ng and Rainboth, 2005)

Image is not available.

Common name(s): No common name for this new species.

Global distribution: Tropical; Southeast Asia: Cambodia.

Ecology: This freshwater, demersal species is found among leaf litter in a fast-flowing stream with a sandy bottom.

Biology: Mouth of this species is terminal. The front of the lower jaw bears the snout's tip. The anterior and posterior nostrils are small. The caudal fin is emarginate or truncated; the posterior edge of pectoral spine is serrated. It attains a maximum length of about 3 cm.

Venomous organs: Its pectoral spine and dorsal spine, which are short and straight, are traumatogenic (Ng and Rainboth, 2005).

Akysis fuliginatus (Ng and Rainboth, 2005)

Image is not available.

Common name(s): Not designated as it is a new species.

Global distribution: Tropical; Southeast Asia: Cambodia.

Ecology: This freshwater, demersal species is usually found among roots in a small stream.

Biology: The body of this species is stout. The mouth is terminal. The tip of the snout is not in front of the lower jaw, or is only located slightly in front of the lower jaw. The anterior and the posterior nostrils are small and the anterior nostril is located at the end of a short tube. The caudal fin is emarginate or truncated. This species reaches a maximum standard length of 2.2 cm.

Venomous organs: Its pectoral spine and dorsal spine, which are short and straight, are traumatogenic (Ng and Rainboth, 2005).

Akysis filifer (Ng and Rainboth, 2005)

Image is not available.

Common name(s): Not designated as it is a new species.

Global distribution: Tropical; Southeast Asia: Cambodia.

Ecology: This freshwater, demersal species is usually seen in the sandy bottoms of muddy lakes with moderate currents.

Biology: The body of this species is elongated and the mouth is subterminal. The tip of its snout is in front of the lower jaw. The nostrils are much closer and are well separated by a nasal barbel. The anterior nostril is present on the surface of the snout. The caudal fin is well forked. This fish reaches a maximum length of 5.0 cm.

Venomous organs: Its pectoral spine and two dorsal spines, which are short and straight, are traumatogenic (Ng and Rainboth, 2005).

Akysis hendricksoni (Alfred, 1966)

Image is not available.

Common name(s): Hendrickson's Catfish.

Global distribution: Tropical; Asia: Malaysia and Thailand.

Ecology: It is a benthopelagic freshwater species inhabiting stagnant backwater with substrate composed of sand, mud, silt, and leaf litter.

Biology: The body of this species is relatively stout. The mouth is terminal; and the tip of the snout is only slightly in front of the lower jaw. A space is seen between the anterior nostril and the base of the nasal barbel. The anterior and posterior nostrils are relatively small and the anterior nostril is present at the end of a short tube. The pectoral spine is with four to six serrate on its posterior edge. The caudal fin is almost truncated. This species reaches a maximum length of 4.9 cm.

Venomous organs: The potential venom delivery system which is the bony spine of this species is Y shaped, and the venom gland is paired and is in the anterior grooves of the spine (Wright, 2012).

Akysis heterurus (Ng, 1996)

Image is not available.

Common name(s): To be designated as it is a newly described species.

Global distribution: Tropical; Asia: It is only known from Eastern Sumatra.

Ecology: It is a benthopelagic freshwater species.

Biology: This species has the following characteristic features, namely, moderately compressed body; truncated caudal fin; and presence of three brown spots on the dorsal side, behind the dorsal fin, oval head, and long nasal barbel. the head is fairly depressed and broad. Its dorsal profile is rising uniformly and is sloping ventrally toward the caudal peduncle. Ventral profile, on the contrary, is sloping dorsally to caudal peduncle. Head has small tubercles and its body is with several tubercles in five to six longitudinal rows on each side. Belly, chest, and ventral surface of the head are slightly yellow with delicate brown spots. The coloration of dorsal surface and sides of head and body is brown. Three irregular light-yellow patches are seen over the top and sides. The adipose fin is generally brown, except at the origin and along the edge. The basal two-thirds of the dorsal fin are brown and the remaining one-third is with brown spots. Brown spots are seen in the pectoral, pelvic, anal, and caudal fins. Barbels and pectoral spines are transversely barred with brown rings.

Venomous organs: The presence of four to six comparatively large serrate in the pectoral spine of this species may be traumatogenic causing injuries (Ng, 1966).

Akysis inermis (Ng and Kottelat, 2000)

Image is not available.

Common name(s): None as it is a new species.

Global distribution: Tropical; Southeast Asia: Southern Laos.

Ecology: It occurs in various habitats such as streams and large rivers with stony substrata and strong currents.

Biology: The body of this species is relatively elongate. The mouth is subterminal and tip of the snout is well in front of the lower jaw. The nostrils are close together and are separated only by the nasal barbel. The anterior nostril is present on the surface of the snout. There is a stout pectoral spine and its posterior edge is without serrate. In alcohol-preserved fishes, the ventral surface of the head is light yellow. The adipose fin is brown. While the two-thirds of basal dorsal fin is brown, the remaining portion is only with brown spots. The pectoral, pelvic, and anal fins are provided with brown spots. The caudal fin is strongly forked and its distal half is brown. The pectoral spines and barbels are barred transversely and are with brown rings. This species attains a maximum length of 3.5 cm.

Venomous organs: Its stout pectoral spine may be harmful and traumatogenic to fishermen (Ng and Kottelat, 2000).

Akysis nitidus (Ng and Rainboth, 2005)

Image is not available.

Common name(s): Not designated as it is a new species.

Global distribution: Tropical; Southeast Asia: Laos.

Ecology: It is a demersal, freshwater species. No other data for this species.

Biology: The body of this species is relatively elongate. The mouth is subterminal. The tip of the snout is in front of the lower jaw. The nostrils are close together and are separated only by the nasal barbel. The anterior nostril is located on the surface of the snout, and is not at the tip of a short tube. The caudal fin is strongly forked. This species reaches a maximum standard length of 4.1 cm.

Venomous organs: Its pectoral spine and dorsal spine, which are short and straight, are traumatogenic (Ng and Rainboth, 2005).

Breitensteinia cessator (Ng and Siebert, 1998) (= *Breitensteinia insignis*)

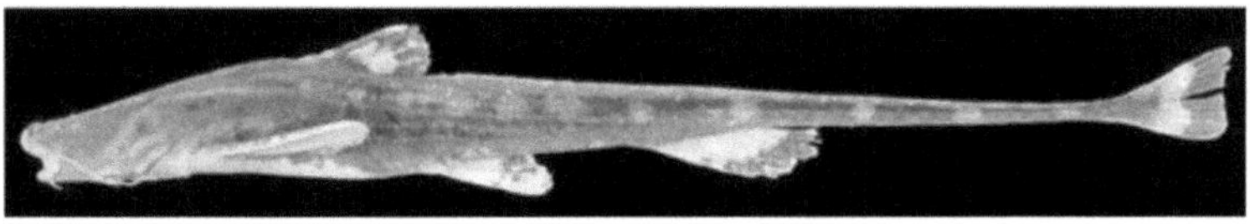

Image credit: Dr. Aque Atanacio (Reproduced with permission).

Common name(s): None.

Global distribution: Tropical; Asia: Indonesia.

Ecology: This demersal species is found among woody debris in the fast-flowing small-to-medium-sized rivers.

Biology: It is a slender fish without an adipose fin but with a long, low adipose ridge. Eyes are large and its caudal peduncle is long and slender. Its head is broad and more depressed. The body is covered with five to six horizontal rows of small, indistinct tubercles on both sides. The anterior nostril is tube-like nostril. The body and head are dark brown in coloration. The dorsum of the body is with scattered light brown spots and its belly which is light brown with few brown spots. The dorsal and pectoral fins are also evenly scattered with brown spots. The caudal fin has a straight posterior margin. The pectoral spine is with 8–16 large serrate in its posterior edge. It attains a total length of 20 cm.

Venomous organs: The venom delivery system, which is the bony spine of this species, is Y shaped. Although the spine is complete, it lacks glandular cells and anterior grooves (Wright, 2012).

Parakysis anomalopteryx (Roberts, 1989)

Image is not available.

Common name(s): Planet Catfish.

Global distribution: Tropical; Southeast Asia: only from West Borneo.

Ecology: Freshwater; this demersal species inhabits shallow waters such as lowland streams among leaf litter.

Biology: This fish is with a highly rugose skin with very small tubercles all over the body. The head is broad and its eyes are small and subcutaneous. The upper lip is without a median lobe and the lower lip is complete. While the branches are absent in the outer mandibular barbel, inner mandibular barbel is with one or two long, slim branches. Both dorsal and pectoral fins have one spine only. The caudal fin which is forked has pointed lobes. It attains a total length of 3 cm.

Venomous organs: Anterior grooves are present (Wright, 2012).

Pseudobagarius inermis (Ng and Kottelat, 2000)

Image is not available.

Common name(s): None.

Global distribution: Tropical; Southeast Asia.

Ecology: This demersal species inhabits fast-flowing small streams and large rivers with stony (gravel to boulder) substrate.

Biology: In this species, its large anterior and posterior nostrils are closely set and are separated only by the base of the nasal barbel. Snout is light yellow with a sharp contrast edge. The caudal fin is deeply forked. The body has a typical color pattern which consists of dorsal and ventral light-yellow patches.

Venomous organs: The potential venom delivery system, which is the bony spine of this species, is Y shaped, and its venom gland is paired and are in the anterior grooves of the spine (Wright, 2012).

Pseudobagarius similis (Ng and Kottelat, 1998)

Image is not available.

Common name(s): Planet Catfish.

Global distribution: Tropical; Asia: Southern Vietnam.

Ecology: It is a benthopelagic freshwater species.

Biology: It has a slender body; an extremely narrow head; and small and closely set eyes. Its pectoral spine has six to eight serrate along its posterior edge. Its caudal fin is strongly forked with pointed lobes. It attains a maximum length of 5.1 cm.

Venomous organs: The venom delivery system, which is the bony spine of this species, is Y shaped, and its venom gland is paired and are in the anterior grooves of the spine (Wright, 2012).

3.2 TORRENT CATFISHES (AMBLYCIPITIDAE)

Amblyceps mangois (Hamilton, 1822)

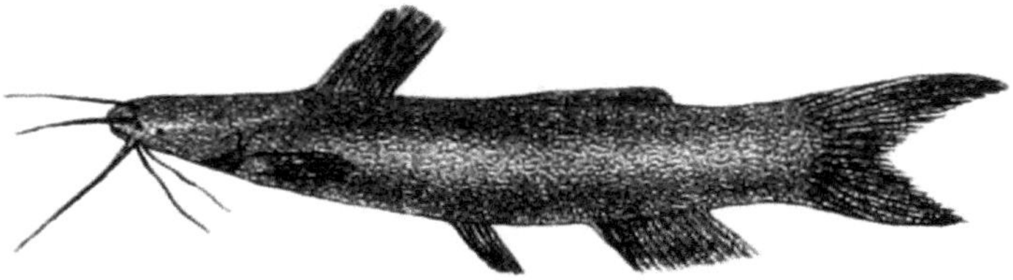

Source: Public domain

Common name(s): Indian Torrent Catfish.

Global distribution: Tropical; Asia: Pakistan to Thailand.

Ecology: This freshwater, benthopelagic species inhabits in pebbly beds of the fast-flowing currents, and exists in bottom of the water with rocks and boulders.

Biology: It has a short body. The eyes are small and subcutaneous. The jaws are unequal as one jaw may be longer than the other. Its teeth of the upper jaw are villiform and are arranged in bands. The lower jaw has a similar narrow band, which is, however, interrupted in the middle. The palate is without teeth. There are four pairs of barbels; maxillary barbels reach middle or end of the pectoral spine that is smooth; nasal barbels equal its head length; and the outer and inner mandibular barbels are as long as the nasal. The caudal fin is deeply forked and its upper lobe is longer. The body coloration is olive brown, lighter below. A dark lateral band is found branching toward the caudal and anal bases. It feeds mainly on aquatic insects. It attains a maximum length of 12.5 cm. It reaches a maximum length of 12.5 cm.

Venomous organs: The venom delivery system, which is the bony spine of this species, is Y shaped and its venom gland is paired, and is in the anterior grooves of the spine (Wright, 2012).

Liobagrus mediadiposalis (Mori, 1936)

Image credit: Dr. Byung–Soo Chae (Reproduced with permission).

Common name(s): South Torrent Catfish.

Global distribution: Subtropical; Southeast Asia: it is endemic to South Korea.

Ecology: This freshwater, benthopelagic species inhabits the rocky or stony bottoms of rocky streams.

Biology: This has a characteristic upper jaw, which is longer and overhanging the lower jaw. Both the jaws are however of equal length. The male has a broader head. A well-developed vascularization is one of the characteristic features of the epidermis of this species. Its dermis lacks scales and consists mostly of a thick, dense connective tissue which is supplied with fine blood capillaries. These features of the skin of this species help in cutaneous respiration. It may, therefore, be able to absorb oxygen through the skin during less oxygenated dry period, that is, summer (Park et al., 2003).

Venomous organs: The potential venom delivery system, which is the bony spine of this species, is Y shaped, and its venom gland is paired and is in the anterior grooves of the spine (Wright, 2012).

Liobagrus reini (Hilgendorf, 1878)

Image credit: Dr. Koichi Goka (Reproduced with permission).

Common name(s): Torrent Catfish.

Global distribution: It is endemic to Japan.

Ecology: It usually lives in the upper to the middle part of rivers with clear water. Its other habitats include cobbles, boulders, bedrock crevices, and roots of vegetation.

Biology: The body of this species is elongated, anteriorly depressed and is posteriorly laterally compressed. The mouth is with four pairs of barbels. The scales are not present on body. The pectoral and dorsal fins are with spinous first ray. The adipose fins are continuous with caudal fin. The posterior margin of the caudal fin is rounded.

Venomous organs: The venom delivery system, which is the bony spine of this species, is Y shaped, and its venom gland is paired and are in the anterior grooves of the spine (Wright, 2012). Dart (2004) reported that the envenomation of this species is common in countries such as Asia, Africa, and Japan.

3.3 MADAGASCAN CATFISHES (ANCHARIIDAE)

Gogo ornatus (Ng and Sparks, 2005)

Image is not available.

Common name(s): Ornate Malagasy Catfish.

Global distribution: Temperate; Africa: East Coast of Madagascar.

Ecology: This freshwater, demersal species is found in well-oxygenated clear river with bottom substrate composed of sand, gravel, and large scattered boulders.

Biology: The body of this species is with highly mottled pigmentation pattern which is with light and dark patches. It has unicuspid teeth on its upper and lower jaws. The eyes are small and it has a straight dorsoposterior margin in its dorsal fin. It reaches a maximum length of 21.2 cm.

Venomous organs: The potential venom delivery system, which is the bony spine of this species, is Y shaped. Multiple bundles of glandular cells are seen along the posterior half of the spine (Wright, 2012).

3.4 SEA CATFISHES (ARIIDAE)

Cathorops agassizii (Eigenmann and Eigenmann, 1888)

Image credit: Rotundo, Matheus M. (Reproduced with permission).

Common name(s): Not designated.

Global distribution: Tropical; South America: Guyana and Brazil.

Ecology: It is a freshwater, benthopelagic species.

Biology: It reaches a maximum length of 32 cm. No other data for this species are available.

Venomous organs and envenomation: It is a traumatogenic species (https://www.fishbase.se/summary/Cathorops-agassizii.html). Junqueira et al. (2006) reported that the stings are often contaminated by Gram-negative bacterial species, namely, *Klebsiella pneumoniae*, *Enterobacter* sp., *Escherichia coli*, *Serratia marcescens*, *Serratia* sp., and *Proteus mirabilis*. This sort of contamination caused environmental public health conditions, as well as, accidents with these fish stings developed significative acute secondary infections in humans.

Cathorops aguadulce (Meek, 1904) (= *Arius aguadulce*)

Source: Public domain

Common name(s): Estuarine Sea Catfish.

Global distribution: Central America: Mexico to Izabal Lake in Guatemala.

Ecology: It is a demersal species inhabits in rivers and lagoons. It may also thrive in brackish waters.

Biology: The head of this species is long and depressed; and snout is long. A shield is seen on the top of the head, which is rough, and is covered with thin skin. A deep, well-defined groove is seen along the center top of head. Its bony process, which is located between head shield and dorsal fin base, is long and narrow behind. There are three pairs of barbells (on chin and both jaws) and the top barbell is reaching past pectoral spine base. The mouth is small and its lower jaws are arched. Its caudal fin is deeply forked. Its lateral line is entire with several branches at the top and bottom base of the caudal fin. The body lacks scales. It attains a maximum length of 29 cm.

Venomous organs and envenomation: The potential venom delivery system, which is the bony spine of this species, is Y shaped. Paired anterior venom glands are present (Wright, 2012). (https://www.fishbase.se/summary/Cathorops-aguadulce) reported that it is a traumatogenic species and the posterior margin of its pectoral-fin spine is with long and conspicuous serrations.

Potamarius nelsoni (Evermann and Goldsborough, 1902)

Image credit: Alexandre P. Marceniuk and Naércio A. Menezes (Reproduced with permission).

Common name (s): Lacandon Sea Catfish.

Global distribution: Tropical; Central America: Mexico.

Ecology: Freshwater; it is a demersal species inhabits only in freshwater systems.

Biology: In this species, males carry the eggs laid by the females in their mouth. It has a maximum length of 39 cm.

Venomous organs: It is a traumatogenic species (https://www.fishbase.in/summary/51770). Its potential venom delivery system, which is the bony spine of this species, is Y shaped. Anterior venom glandular cells are visible (Wright, 2012).

3.5 NAKED SUCKERMOUTH CATFISHES (ASTROBLEPIDAE)

Astroblepus chotae (Regan, 1904)

Image is not available.

Common name(s): Chota Climbing Catfish; Planet Catfish.

Global distribution: Tropical; South America: Magdalena and Orinoco River basins.

Ecology: This is a demersal, freshwater species.

Biology: This species has been reported to attach its eggs individually to the underside of large rocks. It attains a maximum length of 10 cm.

Venomous organs: The potential venom delivery system, which is the bony spine of this species, is N shaped and the glandular cells are not discernible (Wright, 2012).

3.6 DRIFTWOOD CATFISHES (AUCHENIPTERIDAE)

***Ageneiosus* sp.**

Source: Public domain

Common name(s): South American Catfish.

Global distribution: Southern Panama and throughout most regions of South America to Argentina.

Ecology: It is a freshwater species.

Biology: In this unidentified species, maxillary barbel is entirely ossified, forming a rigid, thickened, and elongate barbel, which may reach beyond the anterior eye margin. Two rows of sharp recurved hooks are seen on dorsal margin of barbel. The dorsal-fin spine is elongate, rigid, and aculeiform and is straight to strongly arched. The anterior margin of the dorsal-fin spine is with pungent, unicuspid serrate. It reaches a maximum length of 59 cm.

Venomous organs: The potential venom delivery system, which is the bony spine of this species, is Y shaped and the glandular cells are not discernible (Wright, 2012).

Centromochlus existimatus (Mees, 1974)

Image credit: Luisa Maria Sarmento Soares (Reproduced with permission).

Common name(s): Driftwood catfish.

Global distribution: Tropical; South America: Amazon River basin.

Ecology: It occurs in the main course of the rivers and adjacent lagoons.

Biology: It has an elongate maxilla that extends into the maxillary barbel. Its much-elongated Müllerian ramus disc is projecting deep into the lumen of the swim bladder. There is one spine in the dorsal fin and one in each of the pectoral fin. All these spines are well developed and are sharp with serrated edges (Sazima et al., 2005). It attains a maximum length of 9 cm.

Venomous organ: This species has been reported to have attacked people during bucket bathing. All stings occurred at dusk or early night. The most commonly injured body places were the upper limbs (mainly hands), followed by head and neck and lower limbs (Sazima et al., 2005).

Centromochlus heckelii (De Filippi, 1853).

Image credit: Mark Sabaj (Reproduced with permission).

Common name(s): Driftwood Catfish.

Global distribution: Tropical; South America: Amazon and Orinoco River basins.

Ecology: It occurs in the main course of the rivers and adjacent lagoons.

Biology: It has large eyes and its large pectoral fins are also large and are always spread. Müllerian ramus disc is much elongated, projecting deep into lumen of swim bladder. Its dorsal and pectoral fins are provided with long and sharp serrate spines. This omnivorous fish feeds on insects, microcrustaceans, seeds, fruits, algae, and also on small fish. It is known to attain a maximum length of 15 cm.

Venomous organs and envenomation: Its dorsal and pectoral fins are provided with a well-developed and sharp spine which possesses serrated edges. When frightened, it is spread out and locked readily. This defensive behavior may result in this catfish becoming attached by its spines to several places on a bather's body (Sazima et al., 2005). The most commonly injured body places due to the sting of this fish were the upper limbs (mainly hands), followed by head and neck and lower limbs. Sazima et al. (2005) reported that the serrate spine edges rendered the extraction of the attached fish difficult and painful. Further, as the skin and subcutaneous tissue were pulled during the process, the pain subsided in about half an hour. The puncture wound was found accompanied by local reddening and a slight swelling without further complications.

***Centromochlus* sp.**

Image credit: Dr. Nathan K. Lujan (Reproduced with permission).

Common name(s): Driftwood Catfish.

Global distribution: Tropical; South America: Amazon and Orinoco River basins.

Ecology: It occurs in the main course of the rivers and adjacent lagoons. While smaller fishes take shelter in wood with crevices and gaps, larger fishes are known to occupy in pipes gregariously pipes.

Biology: Not reported.

Venomous organs: The potential venom delivery system, which is the bony spine of this unidentified species, is Y shaped and the glandular cells are not discernible (Wright, 2012).

Entomocorus benjamini (C. H. Eigenmann, 1917)

Image credit: Roberto E. Reis (Reproduced with permission).

Common name(s): Sleeper Catfish.

Global distribution: Tropical; South America: Upper Madeira River basin.

Ecology: It is a freshwater, demersal species.

Biology: In this species, the caudal fin is hyaline, with black pigmented distal half of dorsal lobe and ventral lobe. The pectoral and pelvic fins are, however, unpigmented. It has a shorter anal-fin base. Male fish has an elongated ventral fin and dorsal becomes long and curved. Its main food items are aquatic and terrestrial invertebrates (primarily insects), zooplankton (including cladocerans, copepods, and rotifers), and also on aquatic and terrestrial vegetation. It attains a maximum length of 7 cm.

Venomous organs: The potential venom delivery system, which is the bony spine of this species, is Y shaped and the glandular cells are not discernible (Wright, 2012).

Horabagrus brachysoma (Günther, 1864)

Image credit: Jan Joseph George, Wikipedia. https://creativecommons.org/licenses/by-sa/4.0/

Common name(s): Bullseye Catfish; Sun Catfish; Gunther's Catfish; Yellow Catfish; Solar Catfish; Golden Red Tail Catfish.

Global distribution: It occurs mostly in Indian riverine systems.

Ecology: It inhabits lowland areas of rivers and backwaters with sand or mud substrata; deep pools; and hill streams with much vegetation.

Biology: The body of this fish is yellowish with a black shoulder spot. The head is large with a wide mouth. The eyes are large. The dorsal fin has a hard spine and an adipose fin. There are one pair of nasal, one pair of maxillary (sides of the mouth), and two pairs of mandibular (chin) barbels. It is known to feed on a variety of food items such as crustaceans, mollusks, and fish. Large adults may also consume even terrestrial insects and frogs. It may attain a maximum length of 45 cm.

Venomous organs: The potential venom delivery system, which is the bony spine of this species, is Y shaped. There are paired anterior and posterior venom glands (Wright, 2012).

Trachelyopterus coriaceus (Valenciennes, 1840)

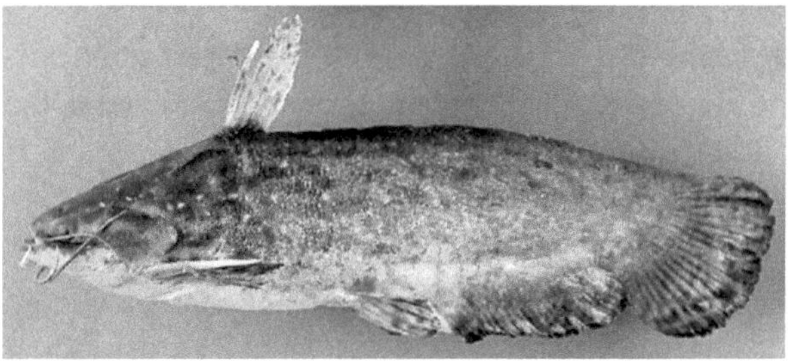

Image credit: Erik Åhlander (Reproduced with permission).

Common name(s): Moors Flue.

Global distribution: Tropical; South America: coastal rivers of the Guianas and Brazil. It is also known from the Amazon River basin.

Ecology: This demersal, freshwater species dwells mainly in swamps or tidal parts of rivers.

Biology: In this species, an adipose fin is absent. It reaches a maximum length of 18 cm. No other information is known.

Venomous organs: The potential venom delivery system, which is the bony spine of this species, is Y shaped and the glandular cells are not discernible (Wright, 2012).

Trachelyopterus striatulus (Steindachner, 1877)
(= *Parauchenipterus striatulus*)

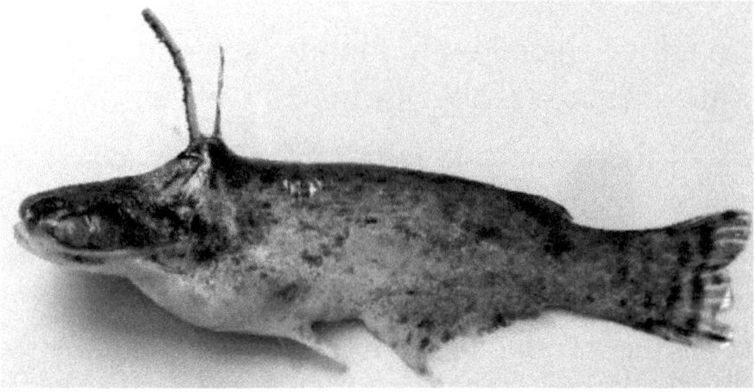

Image credit: Dr. Ezequiel Mabragaña (Reproduced with permission).

Common name(s): Singing Catfish.

Global distribution: Tropical; South America: Coastal rivers in Southeastern Brazil and also in Argentina and Uruguay.

Ecology: This demersal, freshwater species is known to occur mainly in the upper zone of the rivers where lowest transparency prevails.

Biology: This species is characterized by a pair of maxillary barbels. The lips are thin; the upper lip is attached to maxilla and premaxilla; and the lower lip is attached to the mandible. The branchial apparatus of this fish consists of five pairs of gill arches. It attains a maximum length of 28 cm.

Venomous organs: The potential venom delivery system, which is the bony spine of this species, is Y shaped and the glandular cells are not discernible (Wright, 2012).

3.7 BAGRID CATFISHES (BAGRIDAE)

Bagrichthys majusculus (Ng, 2002)

Image credit: Yuichi Kano (Reproduced with permission).

Common name(s): White–Whisker Lancer.

Global distribution: Tropical; Asia: Thailand and Indochina.

Ecology: It inhabits rivers with muddy substrates. It may move into flooded riparian zones to *spawn* during the wet season.

Biology: Body of this species is long and compressed. Its dorsal profile is rising steeply from the tip of snout to origin of the dorsal fin and it is then sloping ventrally to the end of the caudal peduncle. Ventral profile, on the contrary, is flat to the base of the anal fin and then it is sloping dorsally to the end of the caudal peduncle. Both the anus and the urogenital openings are seen vertically through the middle of pelvic fin. Its skin is smooth. The lateral line is complete with long tubular epidermal extensions of the sensory pores. The head is slightly compressed and narrow; its anterior part is produced into a prominent protruding snout. The mouth is inferior and relatively small, with papillate lips. The gill openings are wide and four pairs of barbels are present. Its maxillary barbel is moderately long, extending to opercular flap. The dorsal fin is located at middle of body and its origin is near the tip of snout. The dorsal fin margin is convex. Caudal fin is deeply forked with pointed upper and lower lobes. Alcohol-preserved specimens are brown or dark brown dorsally. A cream-colored mid-lateral streak is running along entire length of body. Large patches of cream coloration may be present on nape. The body and the ventral surfaces of head are also cream colored. It may attain a maximum size of 18 cm in length.

Venomous organs: Its venomous dorsal-fin spine is short, straight, and robust, with 15–29 serrations on its posterior margin. The potential venom delivery system, which is the bony spine of this species, is Y paired and there are anterior and posterior venom glands (Wright, 2012).

Bagrus docmak (Forsskål, 1775)

Source: Public domain

Common name(s): Semutundu.

Global distribution: Africa.

Ecology: This benthopelagic species inhabits lakes, swamps, and rivers; and it is found distributed widely in both shallow and deep-water areas. This species is often associated with rocky bottoms/coarse substrates. It is known to be mostly active during night and twilight hours.

Biology: The body of this species is slightly elongated and it maybe 5–6 times longer than deep. The head is depressed, broad, and is about times as long as wide. There are four pairs of barbels and a long adipose dorsal fin is present. It feeds on insects, crustaceans, mollusks, and fish. Vegetable matter and debris are its major food items. It may attain a maximum length of 127 cm.

Venomous organs: The potential venom delivery system, which is the bony spine of this species, is Y shaped and anterior glandular cells are present (Wright, 2012).

Batasio tengana (Hamilton, 1822) (= *Batasio affinis*)

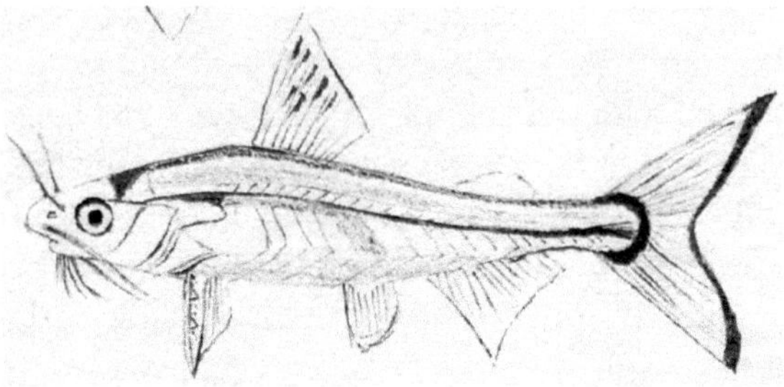

Source: Public domain

Common name(s): Assame Batasio.

Global distribution: Tropical; Asia: India, Bangladesh, Myanmar, and Thailand.

Ecology: Freshwater; this benthopelagic species occurs in rivers, canals, and torrential streams.

Biology: It has a light brown body with five to six vertical dark brown bars. Large sensory pores are seen on the head. Males are provided with a long, slender genital papilla which is extending to base of second anal-fin ray. It may reach a maximum length of 9 cm.

Venomous organs: The potential venom delivery system, which is the bony spine of this species, is Y shaped. Small and paired posterior venom glands are seen (Wright, 2012).

Chandramara chandramara (Hamilton, 1822) (= *Rama chandramara*)

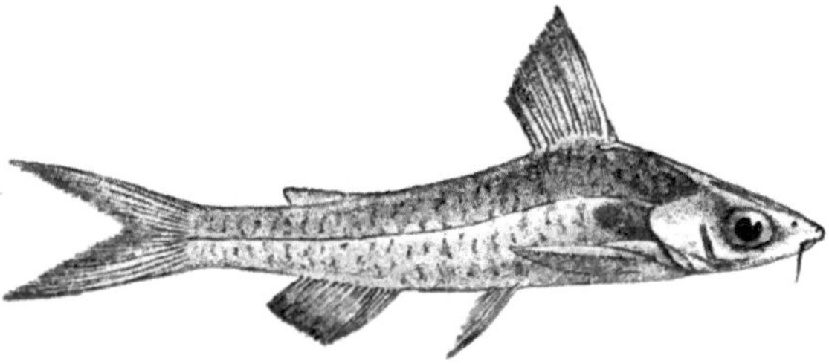

Source: Public domain

Common name(s): Asian Cory; Hummingbird Catfish; Hovering Catfish.

Global distribution: It is common in India and Bangladesh.

Ecology: It dwells in streams, canals, ditches, and beels.

Biology: The body of this species is short and compressed with a rounded abdomen. The head is small and compressed; its snout is obtusely rounded; and the mouth is subterminal and is moderately wide with subequal jaws. The eyes are large and are placed ventrally. The teeth of the jaws and palate look like broad villiform bands. There are four pairs of slender barbels, which are arranged as one each of maxillary and nasal regions and two are of mandibular. The caudal fin is forked with equal lobes. The lateral line is complete with a few pores above pectoral fin base. The body is yellowish with dark to light brown spots. The broad dark brown lateral line is present with two thinner lines above and below. All fins are hyaline with dark first hard rays. It feeds mainly on white worm and tubifex. It reaches a maximum length of 5 cm.

Venomous organs: The potential venom delivery system, which is the bony spine of this species. is Y shaped and there is a single circumferential venom gland (Wright, 2012).

Hemibagrus spilopterus (Ng and Rainboth, 1999)

Image credit: Yuichi Kano (Reproduced with permission).

Common name(s): Black Spotted Catfish.

Global distribution: Tropical; Asia: known only from Cambodia.

Ecology: This demersal fish inhabits slowly flowing or standing waters including main river channels and lakes. It may undergo localized migration into flooded forest during the wet season.

Biology: This species is characterized by its depressed head, a lean occipital process, and a short adipose fin. Its first three dorsal-fin rays lack filamentous extensions. It feeds mainly on aquatic shrimps and other crustaceans; aquatic insects; and fishes. It may attain a maximum length of 31 cm.

Venomous organs: The potential venom delivery system, which is the bony spine of this species, is Y paired and there are anterior and posterior venom glands (Wright, 2012).

Hyalobagrus flavus (Ng and Kottelat, 1998)

Image is not available.

Common name(s): Ornate Shadow Catfish.

Global distribution: Tropical; Asia: Sumatra, Borneo, and Indonesia.

Ecology: This demersal freshwater species dwells in habitats nearer to peat swamps. It may form large shoals and sometimes it may mix with other shoaling species like *Pseudeutropius moolengurghae* and *Pseudeutropius brachypopterus.*

Biology: Its body is mostly transparent and is with a one horizontal black stripe and several black spots. It has a small mouth. The distal half of anterior edge of its pectoral spine is serrated. The dorsal fin is without black margin. Males possess elongated genital papilla and gravid females are with a green-colored belly area. It is known to prey on smaller dwarf shrimp. It attains a maximum length of 4.6 cm.

Venomous organs: The potential venom delivery system, which is the bony spine of this species, is Y shaped. The posterior glandular cells are visible (Wright, 2012).

Mystus mysticetus (Roberts, 1992)

Common name(s): Striped Dwarf Catfish.

Global distribution: Tropical; Asia.

Ecology: This demersal species is potamodromous. It is commonly seen in mixed schools with species such as *Mystus multiradiatus*, which congregate around solid objects. It is also found moving into the flooded forests especially during rainy season.

Biology: In this species, the eyes are lateral, rather than dorsolateral. Its mouth is nearly terminal and is less downturned. Serrate of pectoral spine are smaller and less erect. The adipose fin is fairly short and is originating well behind dorsal fin. Its maxillary barbel is found extending to beyond anal fin posteriorly. There are two whitish and three faint dark stripes on the side of the body. Its anal and caudal fins are black in their tips. Male has a slender genital papilla in front of anal fin. It feeds on zooplankton, crustaceans, and aquatic insects. It reaches a maximum length of 13 cm.

Venomous organs: The potential venom delivery system, which is the bony spine of this species, is Y shaped. The anterior and posterior glandular cells are visible (Wright, 2012).

Nanobagrus nebulosus (Ng and Tan, 1999)

Image is not available.

Common name(s): Not designated.

Global distribution: Tropical; Asia: Malaysia.

Ecology: This benthopelagic species inhabits streams.

Biology: The body coloration of this species is brown. Three rows of cream-colored spots are present above on both sides below the lateral line. The posterior edge of its pectoral spine has 16 serrations. It attains a maximum length of 3.5 cm.

Venomous organs: The potential venom delivery system which is the bony spine of this species is Y shaped and a single circumferential, venom gland is present (Wright, 2012).

Olyra longicaudata (McClelland, 1842)

Image credit: Frank M. Greco (Reproduced with permission).

Common name(s): Asian Fighting Catfish; Bannertail Catfish.

Global distribution: Tropical; Asia: India and Myanmar.

Ecology: This demersal species inhabits small rocky streams at the base of hills.

Biology: The body of this species is long and slender with a reddish pink coloration. It attains a maximum length of 11 cm.

Venomous organs: The potential venom delivery system which is the bony spine of this species is Y shaped. Glandular cells are seen near anterior groove (Wright, 2012).

Pseudobagrus crassilabris (Günther, 1864) (= *Leiocassis crassilabris; Tachysurus crassilabris*)

Pseudobagrus sp.
Source: Public domain

Common name(s): Bagrid Catfish.

Global distribution: Subtropical; Asia: China.

Ecology: It is a demersal species and is inhabiting from fast-flowing hill streams to large rivers.

Biology: The species of this genus are characterized by a pterotic with a broad lateral laminar expansion and an extrascapular without a laterosensory canal. It attains a maximum length of 24 cm.

Venomous organs: The potential venom delivery system, which is the bony spine of this species, is Y shaped. Paired anterior and posterior venom glands are seen (Wright, 2012).

Pseudomystus siamensis (Regan, 1913)

Image credit: Fishing Adventures Thailand (Reproduced with permission).

Common name(s): Asian Bumblebee; Siamese Barred Catfish.

Global distribution: Tropical Asia: Chao Phraya and Mekong basins; Peninsular and Southeast Thailand river systems;

Ecology: This demersal species occurs in rivers and streams.

Biology: The body coloration is either plain dark or yellowish to dark gray with irregular vertical bars. The caudal fin is hyaline and it may or may not have a black blotch on its lobes. The barbels are short and maxillary ones are not reaching base of pectoral spine. The adipose fin is high and rounded. It feeds on aquatic insect larvas including odonatans. Male guards the eggs laid by the female. It reaches a maximum length of 15 cm.

Venomous organs: The potential venom delivery system, which is the bony spine of this species, is Y shaped. The paired venom glands are hemispherical (Wright, 2012).

Sperata aor (Hamilton, 1822)

Image credit: Sufe, Wikimedia. https://creativecommons.org/licenses/by-sa/4.0/

Common name(s): Long Whiskered Catfish; Indian Shovelmouth Catfish.

Global distribution: Tropical Asia: Pakistan, India, Nepal, Myanmar, Bangladesh, and Nepal.

Ecology: This demersal is found in rivers, reservoirs, ponds, lakes, and channels.

Biology: The head of this species is depressed and snout is rounded. Among its four pairs of barbels, the maxillary pair is almost nearer to the end of the anal fin. A deeply forked caudal fin is seen. The body is silvery gray with white below. Its fins are yellowish with dark edges. Black blotch is seen at the posterior end of the adipose fin. This predatory fish feeds on small fishes and worms. It reaches a maximum length of 180 cm.

Venomous organs: The potential venom delivery system, which is the bony spine of this species, is Y shaped. Glandular cells are not visible (Wright, 2012).

3.8 ARMORED CATFISHES (CALLICHTHYIDAE)

Aspidoras taurus (Lima and Britto, 2001)

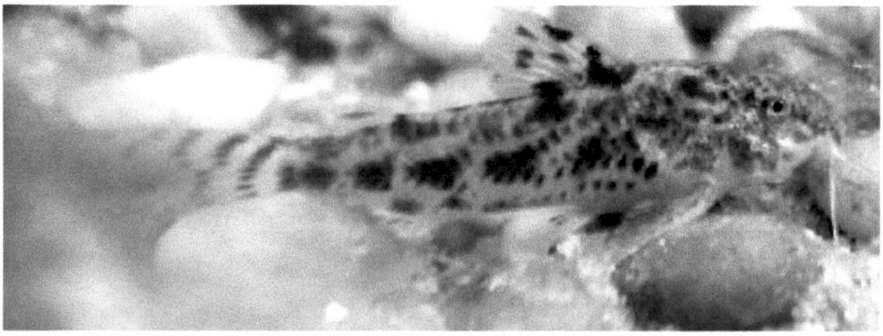

Image credit: Juan Felipe, Wikimedia. https://creativecommons.org/licenses/by/3.0/deed.en

Common name(s): Not reported.

Global distribution: Tropical; South America: Brazil.

Ecology: It is a benthopelagic species.

Biology: In this species, there are two pairs of maxillary barbels and a pair of mental barbels. A small, curved, and bony spine is seen in front of its adipose fin. It attains a maximum length of 5.4 cm.

Venomous organs: The potential venom delivery system, which is the bony spine of this species, is Y shaped and there are circumferential glandular cells (Wright, 2012).

Callichthys callichthys (Linnaeus, 1758)

Image credit: Chucao, Wikipedia. https://creativecommons.org/licenses/by-sa/3.0/

Common name (s): Armored Catfish; Bubblenest Catfish.

Global distribution: Subtropical; South America: Argentina.

Ecology: This demersal species is found in extreme environments from slack water zones with dense vegetation to fairly slightly turbid waters with the influence of free-flowing streams. During adverse conditions, this fish can move out of the water to another aquatic environment due to its intestinal breathing ability.

Biology: Its body is elongate and is tapering in breadth posteriorly. Its head is broad and flattened dorsally. Two rows of bony scutes are seen on the flanks. The caudal fin is rounded. There are two pairs of maxillary barbels which reach to the pectoral fins. In this species, eyes are small. Females are large and robust with a dull olive-green coloration. Conversely, males are with brighter with a violet sheen laterally. Further these males possess a well-developed and longer pectoral-fin spine which is reddish-brown and is edged with reddish orange. This nocturnal fish eats mainly fish, insects and other plant matter. It attains a maximum length of 20 cm.

Venomous organs: The potential venom delivery system, which is the bony spine of this species, is Y shaped and there are circumferential glandular cells (Wright, 2012).

Corydoras aeneus (Gill, 1858)

Source: Public domain

Common name(s): Green Corydoras; Bronze Corydoras Bronze Catfish; Wavy Catfish; Light Spot Corydoras.

Global distribution: It is a common species in South America.

Ecology: It is found both in running waters and in stagnant shallow waters with soft bottoms and in running waters. It may also tolerate heavily polluted waters with poor oxygen levels.

Biology: It has a pink or yellow body with a blue-gray head and back. Ventrum is white in color. Its fins are pink or yellow. Its adipose, dorsal and pectoral fins also possess a sharp barb. A brownish-orange patch is usually seen on its head in front of the dorsal fin. As many other species of catfishes, females of this species are larger with a fairly higher body frame. It reaches a maximum length of 7.5 cm.

Venomous organs: The potential venom delivery system, which is the bony spine of this species, is Y shaped and there are circumferential glandular cells (Wright, 2012). Its sharp barb with mild poison has been reported to serve as sting to harm other fishes.

Corydoras splendens (Castelnau, 1855)

Image credit: Maschinenkanone, Wikipedia. https://creativecommons.org/licenses/by-sa/2.0/de/deed.en

Common name(s): Emerald Catfish.

Global distribution: Tropical; South America: Amazon River basin.

Ecology: This benthic species dwells in quiet waters with dense vegetation or in shallow muddy waters.

Biology: the body of this fish reflects a metallic green, shade of blue green, or even a bluish color depending upon the angle of lighting. While its belly and pectoral, ventral, and anal fins are yellowish, its dorsal, caudal, and adipose fins are translucent brownish. Females are generally larger and more robust than their counterparts and they have a more pinkish belly against more yellowish one in males. Its diet includes small crustaceans, insect larvas, and worms. It attains a total length of 7.5 cm.

Venomous organs: The potential venom delivery system, which is the bony spine of this species, is Y shaped and no glandular cells are visible (Wright, 2012).

Dianema longibarbis (Cope, 1872)

Image credit: Miguel Angel Landinez Parra (Reproduced with permission).

Common name(s): Porthole Catfish.

Global distribution: Tropical; South America: Amazon River basin.

Ecology: This is a demersal species.

Biology: The coloration of the head and body is gray/brown and the coloration of the belly is lighter. The head and body are with scattered black spots, which may vary in number and intensity. All the fins are light tan in color. The caudal fin is slightly forked. It is known to reach a maximum length of 8.2 cm.

Venomous organs: The potential venom delivery system, which is the bony spine of this species, is Y shaped. There are paired posterior glands and a few anterior glandular cells are seen (Wright, 2012).

Hoplosternum littorale (Hancock, 1828)

Image credit: Dr. Miguel Angel Landinez Parra (Reproduced with permission).

Common name(s): Armoured Catfish.

Global distribution: Subtropical; South America: Buenos Aires and Argentina.

Ecology: This demersal species inhabits mainly swamps.

Biology: In this species, males grow to a larger maximum size than females. During the reproductive season, this fish has been reported to develop fat deposits in its pectoral fins which are reddish. During rainy season, this species consumes mainly chironomids associated with detritus; and on terrestrial insects, microcrustaceans, aquatic Diptera, and detritus during the dry season. It is also known absorb a good quantity of anaerobic bacteria from the substrate. It attains a maximum length of 26 cm.

Venomous organs: The potential venom delivery system, which is the bony spine of this species, is Y shaped and there are unorganized anterior glandular cells (Wright, 2012).

3.9 ANGLER CATFISHES (CHACIDAE)

Chaca chaca (Hamilton, 1822)

Source: Public domain

Common name(s): Frogmouth Catfish; Squarehead Catfish.

Global distribution: Tropical; Asia: India, Bangladesh, Nepal, Malaya, Indonesia and Myanmar.

Ecology: This demersal fish inhabits rivers, beels, canals, and ponds. It may, however, exist in substrates with soft mud where it conceals its body.

Biology: The body of this species is tadpole like and it is strongly depressed anteriorly and posteriorly it is strongly compressed. There are no lateral scutes but the body is covered with a thick horny skin. Mouth is very broad. The dorsal fin is small. Its rounded caudal fin extends far forward on dorsal and ventral surfaces. A pair of very short, peg-like barbels is present at the corners of the mouth. The older ones have small, arborescent appendages at the corners of their mouth. The body coloration is blackish brown with numerous black and pale spots and blotches. The head is somewhat paler and belly is white with closely-approximated dark blotches. Fins are dark brown with black blotches. A worm-like appendix is present at the fringe of its mouth and it is believed to attract prey. It attains a maximum length of 20 cm.

Venomous organs: The potential venom delivery system, which is the bony spine of this species, is Y shaped. There are paired anterior glands (Wright, 2012).

3.10 LABYRINTH CATFISHES (CLARIIDAE)

***Clarias* spp.**

Most of the species of *Clarius* are definitely venomous and possess venomous spines in front of their dorsal fins. The venom of these species ranges from similar to a bee-sting to that requiring hospitalization (https://www.myaquariumclub.com/are-clarias-catfish-poisonousi-have-read-online-that-there-are-some-cases-...-507056.html; https://animaldiversity.org/accounts/Clarias_batrachus/)

Clarias batrachus (Linnaeus, 1758)

Source: Public domain

Common name(s): Walking Catfish; Philippine catfish.

Global distribution: Asia: Java, Indonesia.

Ecology: This freshwater, brackish, demersal and potamodromous species dwells in several freshwater habitats such as swamps, lowland streams, ponds, pools, ditches and rice paddies. This species is also often seen in stagnant, muddy water. It is known to tolerate hypoxic, stagnant waters.

Biology: This air breathing catfish has the ability to walk or wriggle across dry land tofind food or favorable environments. For this movement, it uses largely its pectorals. It has an elongated body which is compressed posteriorly. It has a little projecting upper jaw. The body coloration is gray or grayish brown, and white spots are seen laterally. There are several pairs of sensory barbels. Its scaleless skin is covered with mucus, which largely protects this fish when it crawls on land. The genital papilla is elongated and pointed in males. It attains a maximum length of 50 cm and weight of 1.2 kg.

Food and feeding: In the wild, it is omnivorous consuming smaller fish, and invertebrates. However, it is also known to feed on detritus and aquatic weeds.

Reproduction: Male and female often show the "spawning embrace." Male often wraps its body around the female and the female in turn releases a stream of adhesive eggs into an already built nest. In Southeast Asian waters, its spawning period is associated with the flooding of rivers during the rainy season and these fishes are able to form nests in areas like dikes of flooded rice fields or submerged mud banks

Venomology: The spine of its pectoral fins is rough on its outer edge and its inner edge is serrated (https://www.fishbase.se/summary/3054)

Male *Clarias batrachus* has been reported to possess venomous pectoral and dorsal spines and its venom is mild compared to some African species, which can admit the victim in hospital (https://www.myaquariumclub.com/are-clarias-catfish-poisonousi-have-read-online-that-there-are-some-cases-...-507056.html).

Venkaiah and Lakshmipathi (2000) reported on the chemistry of its poisonous spine extracts and epidermal secretions. They found that the epidermal secretions showed 8-fold protein content than its counterpart. Similarly, there was a significant variation in trichloroacetic acid (TCA) soluble peptides between these two tissues. It is also evident that the presence of protein secreting cells around the pectoral spine could have led to the evolution of venomous gland in catfish.

Clarias brachysoma (Günther, 1864)

Image credit: Ton Nientied, Wikimedia Commons. https://creativecommons.org/licenses/by/3.0/

Common name(s): Walking Catfish; Teysmann's Spotted Catfish; Magura.

Global distribution: Asia; and it is endemic to Sri Lanka.

Ecology: This freshwater and demersal species inhabits rivers, tanks, pools with muddy water.

Biology: In this species, its body is laterally compressed and elongated (ribbonlike) and is without scales. The head is more depressed. The snout is rather broad and is with small eyes. Its mouth is terminal and teeth are in villiform bands on jaws. The barbels are in four pairs. The nasal barbel is longer than half length of head to end of occipital process. The pectoral spine is very tough and the last rays of the dorsal and anal fins are found united with the tail fin. Its body color is olive-green brown or yellow with small irregular dark spots as large as its eye. Juveniles are dark brown or copper red in color. It attains a maximum length of 50 cm.

Food and feeding: It is a carnivorous, nocturnal fish feeding mainly on worms, crustaceans, fish, frogs, and tadpoles. It has been reported to possess large stomach, spiral valves and highly developed pyloric.

Reproduction: In this species, spawning occurs in very shallow water in makeshift nests among weeds. Male is believed to guard the nest containing eggs till the released young ones attain a length of about 1 cm.

Venomology: It is a venomous species (https://www.fishbase.se/summary/13394).

Clarias fuscus (Lacepède, 1803)

Image credit: Dr. Kwang–Tsao Shao (Reproduced with permission).

Common name(s): Hong Kong Catfish; Whitespotted Clarias; Chinese Catfish.

Global distribution: Subtropical; Asia: China, Taiwan, Philippines and Vietnam.

Ecology: This demersal fish occurs in streams, reservoir, ditches, and ponds; It, however, exists in deeper pools and hides under thick mats of vegetation.

Biology: This species is nude without scale but is sticky. Its body has dark gray or yellowish-brown coloration. Lateral white dots are seen on the sides. Its long dorsal and anal fins and rounded tail fin are the characteristic features of this species. This nocturnal fish forages on small fishes, worms, crustaceans and insects. It attains a maximum length of 40 cm.

Venomous organs: It is a venomous species (https://www.fishbase.se/summary/Clarias-fuscus).

Clarias gariepinus (Burchell, 1822)

Image credit: W.A. Djatmiko, Wikimedia Commons. https://creativecommons.org/licenses/by-sa/3.0/

Common name(s): North African Catfish.

Global distribution: Subtropical: from Algeria to South Africa; from Nile to West Africa.

Ecology: The adult fishes occur mainly in quiet waters such as lakes and pools. However, they are known to exist fairly in shallow and soft muddy swampy areas. Based on the water turbidity and substrate type, two color types of this species are known.

Biology: It has an elongated body that is smooth and very flexible. Its head is rectangular and pointed dorsally. The snout is broadly rounded. Its eyes are relatively small. Its maxillary barbels are rarely shorter than head and are usually somewhat longer. The outer mandibular barbel pair is longer than the inner pair. The dorsal fin reaches caudal fin and the anal fin origin is closer to the caudal fin base. This long dorsal fin, however, is without a sting. The pectoral spine is strong and is serrated only on its outer side. The lateral line is a small, white line which extends from the rear end of the head to the middle of the tail fin base. It has an organ for breathing with air, which is a gill chamber. This nocturnal species feeds on a variety of foods such as zooplankton, invertebrates and fish. It is also known to consume insects, young birds, rotting flesh, and plants. It reaches a maximum length of 1.7 m and weight of about 60 kg.

Venomous organs: The venom apparatus of this species has a sharp and stout sting which is located right in front of the soft-rayed pectoral fin portion. The presence of strong articulations of the sting help its erection and locking during the envenomation of intruders. The venom glands of this species are organized as masses of single-cell glands which are present in the proximity of the stings (Satora et al., 2008).

Clarias theodorae (Weber, 1897)

Image credit: Penny Haworth (Reproduced with permission).

Common name(s): Snake Catfish.

Global distribution: Tropical; Africa.

Ecology: This demersal species lives in vegetated areas in shallow and swampy places around the lake. It may also be found hiding under rocks.

Biology: The head of this species is rectangular in dorsal outline. Its snout is broadly rounded. The eyes are dorsally located. The tooth plates are relatively small. The dorsal and anal fins of this species are not in line with the tail fin. Its gill rakers, which are distantly set, are fairly long and slender. This species is omnivorous feeding on insects, plant seeds, and other fishes. It attains a maximum length of 35 cm.

Venomous organs: The potential venom delivery system, which is the bony spine of this species, is Y shaped. Paired hemispherical glands are seen (Wright, 2012).

Dinotopterus cunningtoni (Boulenger, 1906)

Image credit: Hubert Szczygiel. North American Native Fishes Association. https://uk.inaturalist.org/taxa/99303-Dinotopterus-cunningtoni

Common name(s): None.

Global distribution: Tropical; Africa: it is endemic to Lake Tanganyika.

Ecology: This demersal species has a depth range of 10–130 m; it is more common in areas with rocky substrates.

Biology: This species has its characteristic adipose fin that is clearly seen. It is omnivorous and predatory. It has a maximum length of 175 cm.

Venomous organs: The potential venom delivery system, which is the bony spine of this species, is Y shaped. Circumferential glandular cells are seen (Wright, 2012).

Encheloclarias velatus (Ng and Tan, 2000)

Image is not available.

Common name(s): None.

Global distribution: Tropical; Eastern Indian Ocean: Sumatra and Indonesia.

Ecology: It is a benthopelagic species.

Biology: The dorsal surface of the head and body of this species are evenly dark gray and ventrum is pale. The adipose fin which is not supported by extensions of the neural spines is very dark gray with a hyaline margin. Conversely, the fin rays of other fins are dusky and are with hyaline inter-radial membranes. The anal and the caudal fins are confluent only at the base. It attains a maximum length of 17 cm.

Venomous organs: The potential venom delivery system, which is the bony spine of this species, is Y shaped. The spine is tripped and there are no visible glandular cells (Wright, 2012).

Gymnallabes typus (Günther, 1867)

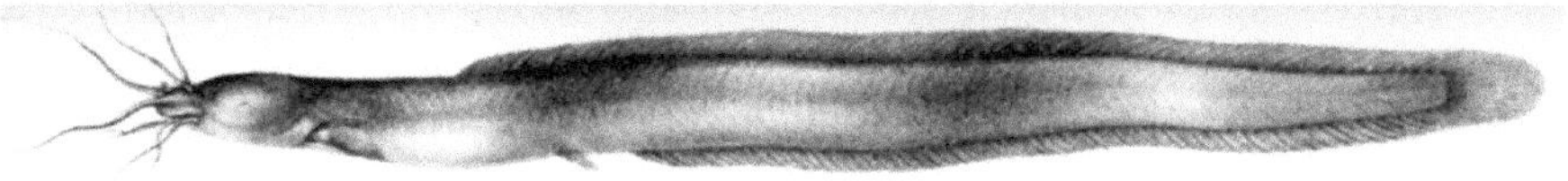

Source: Public domain

Common name(s): Flathead Eel Catfish.

Global distribution: Tropical; Africa: Niger River to Cameroon.

Ecology: This benthic species is often seen in ditches and small streams. It may also be found in swamp regions.

Biology: The body of this species is very elongated and the head is very short. The pectoral fins are uniformly light brown. The dorsal and the anal fins are confluent with caudal fin. Its food items include insect larvas and worms. It attains a maximum length of 30 cm.

Venomous organs: The potential venom delivery system, which is the bony spine of this species, is Y shaped. Circumferential glandular cells are seen (Wright, 2012).

Heterobranchus longifilis (Valenciennes, 1840)

Source: Public domain

Common name(s): Vundu; Vundu-malle; Sampa.

Global distribution: It is largest freshwater species in Southern Africa.

Ecology: It occurs commonly in large deep rivers or in deep pools and lakes.

Biology: The head of this species is large. The barbels are very long and are reaching beyond head. Its rayed dorsal fin reaches above anterior base of anal. The adipose fin is large, extending from rayed dorsal to caudal. The anal fin origin is nearer caudal fin base. Fewer gill rakers are present on the first branchial arch. It is most active at night and is known to feed on invertebrates and insects when small and on fish and other small vertebrates when it is large and grown. It may also scavenge off large carcasses and offal from riverside villages. It may attain a maximum length of 61 cm.

Venomous organs: The potential venom delivery system, which is the bony spine of this species, is Y shaped. Spine is broken and glandular cells are present (Wright, 2012).

Tanganikallabes alboperca (Wright and Bailey 2012)

Image is not available.

Common name(s): None as it is a new species.

Global distribution: Tropical; Africa: Lake Tanganyika.

Ecology: This demersal and benthopelagic species may inhabit rocky bottoms at a depth range of 2–3 m.

Biology: Its body skin is thick with a number of vertical ridges and folds. Its head is depressed and broad. The snout is short with bluntly rounded margin. Its mouth is terminal; the lips are narrow and papillate; and the jaws may be equal, or the upper jaw may be fairly longer than the lower jaw. The nasal barbel is short; the maxillary barbel may extend beyond margin of opercular flap; the lateral mandibular barbel extends to lower opercular margin; and the medial mandibular barbel is short. All the barbels are smooth. The dorsal fin is elongated without spine. The adipose fin is absent. Anal fin is elongated and caudal is rounded. In alcohol-preserved specimens, the dorsum and flanks are uniformly light to dark brown and belly is lighter. All barbels are pigmented, and the fins possess narrow, depigmented border, and brownish bases. In its stomach, fish eggs, crab, and insect larvas were found as food items. It attains a maximum length of 16 cm (Wright and Bailey, 2012).

Venomous organs: Pectoral fin is with a strong spine which has well-developed venom glands. The pectoral-fin spine is only half the size of the pectoral fin and its posterior margin has a maximum of five very small, retrorse serrations (Wright and bailey, 2012).

Tanganikallabes mortiauxi (Poll, 1943)

Image is not available.

Common name(s): None as it is a new species.

Global distribution: Tropical; Africa: Lake Tanganyika.

Ecology: This demersal and benthopelagic species occurs in rocky bottoms of littoral to benthic zones at depth of 2–3 m.

Biology: The body of this species is elongated and is moderately compressed in front of the dorsal fin. The predorsal profile is slightly convex with small indentation. The prepelvic profile is convex. Skin on body forms numerous vertical ridges and folds which are encasing all fins. The head is depressed and broad. The eyes are small and are located dorsolaterally. Its mouth is terminal, lips are narrow and papillate, and jaws are equal or upper jaw may be slightly longer. Mandibular, premaxillary, and vomerine teeth are pointed. All the barbels are covered with thick, tuberculate skin. The dorsal fin is elongate without spine. The adipose fin is absent. The anal fin is elongate. The caudal fin is rounded. In alcohol-preserved specimens, dorsum and flanks are uniformly dark brown to black and belly is fairly lighter. Maxillary, nasal, and lateral mandibular barbels are uniformly dark brown to black. It is an opportunistic and generalist predator feeding on eggs of fishes, crabs, shrimps, insect larvas, etc. It reaches a maximum length of 22 cm (Wright and Bailey, 2012).

Venomous organs: The pectoral fin is with a strong spine which has well-developed venom glands. The pectoral-fin spine is about half the length of the pectoral fin and its posterior margin of spine a maximum of five very small, retrorse serrations (Wright and bailey, 2012).

Tanganikallabes stewarti (Wright and Bailey, 2012)

Image is not available.

Common name(s): None as it is a new species.

Global distribution: Lake Tanganyika.

Ecology: This benthopelagic species may inhabit rocky bottoms at a depth range of 2–3 m.

Biology: The body of this species is elongate and is moderately compressed posterior to origin of dorsal fin. The predorsal profile is convex, with distinct hump formed between origin of dorsal fin. The prepelvic profile is slightly convex. The skin on the body forms numerous vertical ridges and folds. Its head is depressed and broad. Snout is short, with bluntly rounded margin when viewed dorsally. The eyes are small and are located dorso-laterally. Its mouth is terminal; lips are narrow and papillate; and jaws are equal, or upper jaw may be slightly longer. Mandibular, premaxillary, and vomerine teeth are pointed. The nasal barbel is short and is not extending to opercular flap. Its maxillary and lateral mandibular barbels extend to tip of adpressed pectoral-fin spine. The medial mandibular barbel extends slightly beyond lower opercular margin. All barbels are smooth, with very narrow basal membrane. The dorsal fin is without spine. The adipose fin is absent. The anal fin is elongated and the caudal fin is rounded. The dorsum and flanks are brown to dark brown and belly is slightly lighter. The maxillary barbels are uniformly brown. Its nasal and mandibular barbels are brown on proximal half and lighter distally. All rayed fins are uniformly brown. It feeds mainly on fish eggs. It attains a maximum length of 17 cm (Wright and Bailey, 2012).

Venomous organs: The pectoral fin is with a strong spine which has well developed venom glands. The pectoral-fin spine is about half the length of pectoral fin and its posterior margin of spine a maximum of five very small, retrorse serrations (Wright and bailey, 2012).

Tanganikallabes **sp.**

Envenomation: The potential venom delivery system, which is the bony spine of this species, is Y shaped. A single large, circumferential gland is present (Wright, 2012).

Xenoclarias eupogon (Norman, 1928) (= *Xenoclarias holobranchus*)

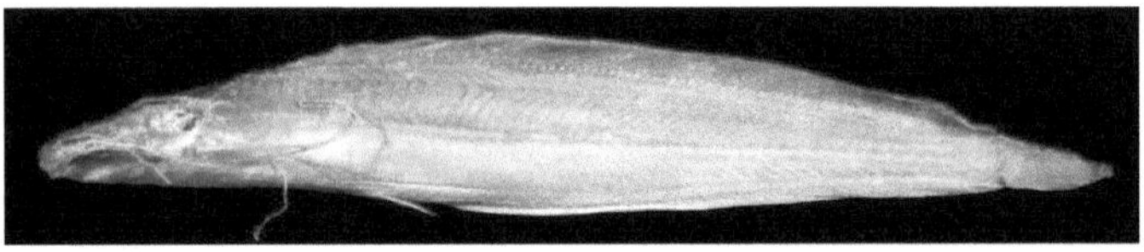

Image credit: The Trustees of the Natural History Museum, London. https://creativecommons.org/licenses/by/4.0/

Common name(s): Lake Victoria Deepwater Catfish.

Global distribution: Tropical Africa: endemic to Lake Victoria.

Ecology: This demersal species inhabits muddy bottoms at a depth range of 10–79 m.

Biology: In this species, females are much broader in the body than males. It is an insectivorous species feeding mainly on *Chaoboridae* larvas. It attains a maximum length of 21 cm.

Venomous organs: The potential venom delivery system, which is the bony spine of this species, is Y shaped. A ventral hemispherical gland visible is seen in the spine (Wright, 2012).

3.11 CLAROTEID CATFISHES (CLAROTEIDAE)

Auchenoglanis occidentalis (Valenciennes, 1840)

Image credit: Linė1, Wikipedia. https://creativecommons.org/licenses/by-sa/3.0/deed.en

Common name(s): Bubu; Giraffe Catfish; African Giraffe Gatfish; Giraffe Catfish; African Giraffe Bagrid.

Global distribution: Tropical; Africa: West Africa; Egypt to Kenya.

Ecology: This demersal fish inhabits shallow lakes and large rivers with muddy bottom.

Biology: The head of this species is large with heavily built and uniformly colored body. Its giraffe-like pattern will fade with age to a two-tone mottled brown, while its belly is creamy white, fins are spottled. The upper jaw has oval-shaped teeth. Hind margin of the adipose fin is rounded; its maxillary barbel, which is shorter than outer mandibular barbel, is generally blackish and it is not reaching beyond eye and much. It is omnivorous feeding on plankton, mollusks, seeds and detritus. It attains a maximum length of 90 cm.

Venomous organs: The potential venom delivery system, which is the bony spine of this species, is Y shaped. Its posterior glands are paired. A small anterior gland is also visible (Wright, 2012).

Bathybagrus tetranema (Bailey and Stewart, 1984)

Image credit: Drs. Reeve M. Bailey and Donald J. Stewart (Reproduced with permission).

Common name(s): Claroteid Catfish.

Global distribution: Tropical; only Lake Tanganyika of Africa.

Ecology: It is a demersal species inhabiting lakes with a depth range of 40–80 m.

Biology: The mouth of this species is subterminal with a longer upper jaw. Its anterior nostrils are located close behind upper lip and they are opening forward on the snout. Its four barbels are thread-like and short; maxillary pair does not extend to the level of the pectoral fin origin; and mental barbels are not quite so far. The dorsal spine is generally short and weak, without serrations on its anterior or posterior edge. The pectoral spine is entire anteriorly with six or seven strong retrorse serrations on its posterior edge; distal serrate are longest and their length is equal to diameter of shaft; and its spine tip is flexible. The anal fin is with five to six simple and seven to nine branched rays. The upper posterior corner of the adipose fin is pointed. The caudal fin is forked. In alcohol-preserved specimens, the body and the upper part of the head are uniformly dusky gray to grayish brown on the top and the sides of the head are covered with small pits which may appear as faint speckles. The lower lip and the underside of the head are sprinkled with melanophores. The dorsal, pectoral, pelvic, and anal fins possess gray speckles; and they are darker basally than distally.

The adipose fin is rather dark, generally with a speckle of white and sprinkled with melanophores. The caudal fin is dusky. It attains a maximum length of 17 cm.

Venomous organs: The potential venom delivery system, which is the bony spine of this species, is Y shaped. The posterior glands are visible (Wright, 2012).

Chrysichthys mabusi (Boulenger, 1905)

Image is not available.

Common name(s): Claroteine Catfish.

Global distribution: Tropical; Africa: Congo and Zambia.

Ecology: It is a freshwater, demersal species.

Biology: In this species, the eye is free from orbital rim. The caudal fin is forked. Mesocoracoid arch is completely ossified. It attains a maximum length of 41 cm.

Venomous organs: The potential venom delivery system, which is the bony spine of this species, is Y shaped. The posterior glands are visible (Wright, 2012).

Clarotes laticeps (Ruppell, 1829)

Source: Public domain

Common name(s): Widehead Catfish.

Global distribution: Tropical; Africa: Chad, Benue, Niger, Senegal, and Volta; also in East Africa and Nile.

Ecology: It inhabits rivers.

Biology: The head of this species is much depressed; its mouth is inferior and the snout is broad. There are four pairs of barbels. The dorsal-fin spine is strong with granulations in front. The adipose fin is rayed and is supported by a spine. Caudal fin is deeply forked and its lobes are marked by a dark band and are rounded. The color of the head and body is olive to dark brown and belly is of silvery white. A blackish blotch is seen above the humeral process. The caudal fin has a blackish or brown band in the middle of each lobe. Other fins are yellowish to light brown. It is known to feed mainly on crustaceans, mollusks, fish, and insects. It attains a maximum length of 80 cm.

Venomous organs: The potential venom delivery system, which is the bony spine of this species, is Y shaped. Paired, anterior glands are visible; and the posterior portion of spine is stripped (Wright, 2012).

Lophiobagrus cyclurus (Worthington and Ricardo, 1937)

Image is not available.

Common name(s): African Bullhead.

Global distribution: Tropical; Africa: Lake Tanganyika.

Ecology: This species is nocturnal and is often hiding in rocks during daytime.

Biology: The body of this species is short, and weekly compressed. The eyes are small and subcutaneous. The body coloration is brown or dusty black. The barbels however, are white. It feeds mainly on small crustaceans, chironomid larvas, and beetle larva. It attains a maximum length of 8 cm.

Venomous organs: The potential venom delivery system, which is the bony spine of this species, is Y shaped. Large, hemispherical glands are evident (Wright, 2012). The mucus secretion of this species is also believed to be toxic to other fishes (https://en.wikipedia.org/wiki/African_bullhead).

3.12 ARMORHEAD CATFISHES (CRANOGLANIDIDAE)

Cranoglanis henrici (Vaillant, 1893)

Cranoglanis sp.

Image credit: Catxx, Wikipedia. https://creativecommons.org/licenses/by-sa/3.0/

Common name(s): Cranoglanidid Catfish; Armorhead Catfish.

Global distribution: Tropical; Asia: Vietnam, and China.

Ecology: It is a benthopelagic species.

Biology: The body of this species is laterally compressed; its head and snout are long and depressed. The dorsal profile rises fairly steeply from tip of snout to dorsal spine, then it descends gently to posterior end of caudal peduncle. The anterior nostril is at tip of short, wide tube and posterior nostril is located directly posterior to anterior nostril. Its mouth is subterminal; and the jaw teeth are villiform. The eyes are moderately large and are laterally situated. The dorsal fin is with spine and 6 rays; and dorsal spine is with 3–6 serrations on anterior edge and 8–14 serrations on its posterior edge. The adipose fin is with short base and rounded posterior margin. Its anal fin is moderately long and the caudal fin is forked with equal lobes. The pectoral fin is with spine and three to four serrations on its anterior edge and about 10 serrations on posterior edge. The body coloration is silvery to gray, darker on the back and white on belly. Fins are brownish with reddish hue and anal margin is darker. It attains a maximum length of 29 cm.

Venomous organs: The potential venom delivery system, which is the bony spine of this species, is Y shaped. Lateral gland remnants are visible (Wright, 2012).

3.13 THORNY CATFISHES (DORADIDAE)

Acanthodoras spinosissimus (Eigenmann and Eigenmann, 1888)

Image credit: Clinton and Charles Robertson, Wikimedia. https://creativecommons.org/licenses/by/2.0/

Common name(s): Chocolate Talking Catfish; Talking Catfish.

Global distribution: South America: Amazon River.

Ecology: This benthic species inhabits standing and slow-moving waters including tributaries; freshwater swamps; backwaters; and coastal mangroves. This nocturnal feeder has also been reported to find refuge among submerged vegetation during day time.

Biology: The head of this fish is wider compared to length. Its body is completely assembled with spines which cover more than half of the body. A thin yellow band is present in its long flank and a dull band is present from the eyes to the beginning of the dorsal fin. All the fins possess markings without a specific pattern. It reaches a maximum length of 15 cm.

Food and feeding: This species is an omnivore feeding largely on bloodworm, tubifex, mosquito larvas, etc.

Reproduction: In this species, both male and female jointly prepare a shallow pit in which the female lays the eggs. These eggs are also guarded by both fish. The eggs are believed to hatch after a period of 4–5 days.

Venomology: It is a venomous species producing toxic fluids (https://www.fishbase.de/summary/acanthodoras-spinosissimus.html). The species of *Acanthodoras* have been reported to secrete a milky toxic substance by an axillary pore situated near its cleithral process. The histological examination, however, showed the presence of an axillary gland with secretory cells Sodre et al. (2021).

Centrochir crocodili (Humboldt, 1821)

Image credit: Flickr, Cornell University Museum of Vertebrates (Reproduced with permission).

Common name(s): Not reported.

Global distribution: South America.

Ecology: It is a freshwater and demersal species.

Biology: It is known to attain a maximum length of 23 cm and weight of 248 g.

Venomous organs: It is a venomous species and it has caused human envenomations especially in South America (Dart, 2004).

Doras micropoeus (Eigenmann, 1912)

Image credit: Erling Holm, Royal Ontario Museum (Reproduced with permission).

Common name(s): Thorny Catfish.

Global distribution: Tropical; South America: Amazon, Essequibo, Demerara, and Corantijn rivers.

Ecology: It is a demersal species inhabiting both lowland floodplain lakes and waterways.

Biology: Head of this species is large, deep, and weakly compressed with an elongate, conical snout (particularly in adults). The body of this species is elongate and is slightly compressed. The ventral surface is flattened from the snout to anal-fin origin. The dorsal profile is concave from snout tip to nares. Its eye is large. The mouth is small and subterminal. Its teeth are present on dentary and premaxilla. Among its three pairs of barbels, maxillary barbel is long and the mental barbels are nearly equal in size. Its lips are fleshy. Its skin is relatively smooth except for minute punctate tubercles which are scattered on its head, body, and fins. The lateral line is surrounded by complete or incomplete series of mid-lateral bony scutes. Dark pigment is more concentrated in distal portion of dorsal fin. It attains a maximum length of 35 cm.

Venomous organs: The potential venom delivery system, which is the bony spine of this species. is Y shaped. Glandular tissue is seen between posterior serrae (Wright, 2012).

Leptodoras nelsoni (Sabaj Pérez, 2005)

Image credit: Mark Sabaj, Academy of Natural Sciences of Philadelphia (Reproduced with permission).

Common name(s): Thorny Catfish.

Global distribution: Tropical; South America: Orinoco River basin.

Ecology: It is a demersal species.

Biology: It has a longer body with longer anal fin. This species has an infranuchal scute, which is the first among its highly developed scutes located mid-laterally Further, the barbels of this fish are fimbriate. It attains a maximum length of 16 cm.

Venomous organs: The potential venom delivery system, which is the bony spine of this species, is Y shaped. Glandular tissue is present between posterior serrae (Wright, 2012).

Lithodoras dorsalis (Valenciennes, 1840)

Source: Public domain

Common name(s): Rock-bacu.

Global distribution: Tropical; South America: Amazon River basin and French Guiana.

Ecology: This is a demersal species.

Biology: The body of this species is elongated, without scales and is covered with bony plaque. Its terminal mouth has three pairs of barbels. The lower jaw is shorter than the upper jaw. Its eyes are small. It feeds on blades of macrophytes and on fruits when the forests are flooded.

Venomous organs: The potential venom delivery system, which is the bony spine of this species, is Y shaped. Glandular tissue is present between posterior serrae (Wright, 2012).

Nemadoras humeralis (Kner, 1855)(= *Opsodoras humeralis*)

Image credit: Jose Birindelli (Reproduced with permission).

Common name(s): Thorny Catfish.

Global distribution: Tropical; South America: native to Amazon River basin; Bolivia, Brazil, Colombia and Peru.

Ecology: It is a demersal species

Biology: The body of this species relatively short and deeply lenticular and is becoming more compressed posteriorly. Its ventral surface is flattened from mouth to anal-fin origin. The head is moderately elongate and is weakly compressed. The dorsal profile is convex from snout tip to dorsal fin origin; and snout is moderately pointed. Its eyes are small with poorly developed adipose eyelid. Its mouth is subterminal to ventral. There are three pairs of barbels, namely, maxillary and inner and outer mental. The maxillary barbel is short and its shaft is with few simple primary fimbriae. Its skin is smooth and is without conspicuous tubercles. The skin pores are more rounded and are in two or three rows. The dorsal surfaces and upper sides of head and body of live specimens are light olive to yellowish green and milky white ventrally. The dorsal fin is dusky along its base with melanophores. The adipose fin is dusky with scattered melanophores. It has a maximum length of 16 cm.

Venomous organs: The potential venom delivery system, which is the bony spine of this species, is Y shaped. Glandular tissue is present between posterior serrae (Wright, 2012).

Physopyxis lyra (Cope, 1872)

Source: Public domain

Common name(s): Thorny Catfish.

Global distribution: Tropical; South America: Amazon and Essequibo River basins.

Ecology: This demersal fish usually inhabits rivers with turbid water where rich organic debris is abundant.

Biology: It has a strong scapular girdle and a single row of spines on its lateral plates and. the dorsal and pectoral spines are strongly ossified. The lateral sides of its dorsal spine possess longitudinal grooves and the basal portion of the spine's anterior margin is serrated. However, the posterior margin is smooth. The pectoral spine is well developed, depressed, and curved; its tip is usually reaching the anal-fin origin. The color of its body is tan with spots and brown blotches. Head, on the contrary, is more pigmented. On its dorsal side, there are 3–4 dark brown saddles that are irregular and are extending on its sides as partial or full pairs. The barbels are tan with brown transverse bands in color. Its fins are provided with dark transverse bands. The tips of the pectoral and the dorsal spines are without pigmentation. Rays and spines are provided with brown transverse bands. Its ventral surface is light or dark pigmented. It has a maximum length of 3.5 cm.

Venomous organs: The potential venom delivery system, which is the bony spine of this species, is Y shaped. Spine is stripped and there are no visible glandular cells (Wright, 2012).

Platydoras armatulus (Valenciennes, 1840)

Image credit: Piotr Kuczynski, Wikipedia. https://creativecommons.org/licenses/by-sa/3.0/

Common name(s): Southern Striped Raphael; Chocolate Catfish; Striped Raphael Catfish.

Global distribution: Subtropical; South America: Paraná River basin.

Ecology: In the wild, this species burrows into the soft, sandy bottom of the rivers during day time and it is a nocturnal fish feeding during night hours.

Biology: It has a long body which is shaped like a tadpole with thick white and black stripes. Males are darker in color, and females are more of a cream color. It feeds on crustaceans, mollusks, and organic debris. It is extremely hardy, having a lifespan of about 15 years. It attains a maximum length of 43 cm.

Venomous organs: It is an armored catfish with several spines which run along its entire lateral line. Its dorsal and pectoral fins are robust and dagger-like ones. Extreme care has to be exercised while handling or netting this fish to prevent envenomation or injury.

(https://www.aquariadise.com/raphael-catfish/; Anon., https://en.wikipedia.org/wiki/Striped_Raphael_catfish).

Platydoras hancockii (Valenciennes, 1840) (= *Amblydoras hancockii*)

Image Credit: Dr. Jan Hvizdak (Reproduced with permission).

Common name(s): Hancock's Catfish; Croaking Spiny Catfish; Talking Catfish.

Global distribution: South America; Amazon basin.

Ecology: This demersal species occurs in stagnant waters or in areas with slow current pattern.

Biology: It has a flattened body with three pair of large barbels which are extending from the mouth. The dorsal fin is fairly tall with a row of spines running from the gills to caudal fin. The body coloration varies from brown to gray with thick or faint markings. It has a maximum length of 10 cm.

Venomous organs: The potential venom delivery system, which is the bony spine of this species, is Y shaped. Glandular tissue is present between posterior serrae (Wright, 2012).

Pterodoras granulosus (Valenciennes, 1821)

Source: Public domain

Common name(s): Granulated Catfish.

Global distribution: Subtropical; South America: Guyana and Suriname.

Ecology: This demersal species which occurs in small groups lives in riverine systems. It is a migratory species travelling between downstream and upstream during March–July and October–January.

Biology: It has a scaleless body but its skin is thick and tough. The body coloration of this species may vary depending upon the location albeit it usually has a muddy-brown color with some darker spots over the body and fins. A total of 23–28 shallow lateral plates known as scutes are seen all along the entire length of the body. Its eyes are very small and there are three simple pairs of barbels. The caudal fin is deeply forked caudal fin. It is an opportunistic omnivore and nocturnal feeder feeding on vascular plants, mollusks, and other food items depending on their availability. This species grows to maximum length of 70 cm and attains a weight of 6.5 kg.

Venomous organs: It is a venomous species and has caused human envenomations in South America (Dart, 2004).

Trachydoras paraguayensis (Eigenmann and Ward, 1907)

Image credit: Dr. Diego Garcia (Reproduced with permission).

Common name(s): Hard-Nosed Thorny Catfish.

Global distribution: Tropical; South America: Paraná River basin.

Ecology: It is a demersal species.

Biology: It is a moderately-sized species with deep body and its maximum depth is at dorsal fin origin. In its lateral view, snout is bluntly rounded. The dorsal profile is rising steeply to the dorsal-fin origin and ventral profile; on the contrary, is relatively flat from gape to anal-fin origin. Its eye is subcircular; mouth is ventral; and gape is small. The upper and lower jaws are edentulous. There are three pairs of barbels. While maxillary barbel is fimbriate and short, mental barbels are short. Gas bladder is slightly large and cordiform. The dorsal half of head and body are yellowish green; dorsal fin rays and pelvic fins are more yellowish than remaining fins. It has a maximum length of 19 cm.

Venomous organs: The potential venom delivery system, which is the bony spine of this species, is Y shaped. Glandular tissue is present between posterior serrae (Wright, 2012).

3.14 SHRIMP CATFISHES (HEPTAPTERIDAE)

Brachyrhamdia heteropleura (Eigenmann, 1912)

Image is not available.

Brachyrhamdia marthae (Sands and Black, 1985)

Source: Public domain

Common name(s): Three-Barbeled Catfishes.

Global distribution: Neotropical; South America: Negro River basins.

Ecology: These are demersal species.

Biology: These are small sized fishes with remarkable color patterns. Its pectoral-fin spines are stiffened and their anterior and posterior margins are provided with ornamentation. It attains a maximum length of 43 mm.

Venomous organs: In these species, the venom glands are associated with the pectoral fin. These venom glands include both surface glands of the ossified pectoral-fin spine and axillary glands the secretions of which are toxic (Carvalho et al., 2021). The cells of the pectoral-fin spine glands are found concentrated near the posterior serrae at the posterior margin of spine. The description of pectoral fin spine and surface glands; and axillary glands of *Brachyrhamdia* is given in the following.

Pectoral-fin spine and surface glands: The pectoral-fin ray of these species is unbranched and stiffened. The pectoral-fin spines have venom gland cells which are concentrated at the posterior margin of the spine, near the posterior pectoral-fin spine serrae (Carvalho et al., 2021).

Axillary glands: The axillary glands of both *Brachyrhamdia marthae* and *Brachyrhamdia heteropleura* are piriform and each axillary gland is divided into many lobules. The values of longest diameter of the axillary gland of *Brachyrhamdia marthae* and *Brachyrhamdia heteropleura* have been reported as 1.3 and 2.94 mm, respectively (Carvalho et al., 2021).

Imparfinis lineatus (Bussing, 1970)

Image is not available.

Common name(s): Three-Barbeled Catfish.

Global distribution: Tropical; Central America: Costa Rica.

Ecology: This demersal species inhabits creeks of moderate currents with velocity between sand, gravel or mud substrata.

Biology: It has a pronounced lateral stripe which is extending from tip of snout to caudal base. It feeds on aquatic insects particularly dipterans and is attaining a total length of 7 cm.

Venomous organs: The potential venom delivery system which is the bony spine of this species is N shaped. Posterior glandular cells are visible (Wright, 2012).

Myoglanis aspredinoides (Donascimiento and Lundberg, 2005)

Image is not available.

Common name(s): None as it is a new species.

Global distribution: Río Ventuari, upper Orinoco in Venezuela.

Ecology: It occurs in stagnant pools of sandy and muddy bottom at a maximum depth of 2 m.

Biology: the body of this species is well elongated, depressed before dorsal fin, and more compressed at caudal. The dorsal profile is convex and ventral profile of head is slightly convex. Its head is about 1.5 times longer than wide and is dorsally covered by thick skin. Its snout is long and wide. The mouth is terminal in position and it has fleshy lips. Its mouth corners extend posteriorly along sides of head. The upper jaw projects slightly beyond the lower jaw. Its maxillary and outer mental barbels are short. The inner mental barbel gets inserted. The eye is ovoid and small and is completely dorsal and protruding. The adipose fin is elongate and low. Anal fin is also elongate. Caudal fin is lanceolate. The pectoral fin is with a strong sharp spine which has three to five serrae and erect dentations along the distal portion of its anterior margin and proximal portion of its posterior margin, respectively. It attains a total length of 12 cm.

Venomous organs: Its pectoral fin that is with a strong sharp spine may be venomous. The envenomation of this new species is yet to be known (Donascimiento and Lundberg, 2005).

***Myoglanis* sp.**

Image credit: Donald C. Taphorn B (Reproduced with permission).

Venomous organs: The potential venom delivery system, which is the bony spine of this species, is Y shaped. A single large, circumferential gland is present (Wright, 2012).

Pimelodella mucosa (Eigenmann and Ward, 1907)

Image credit: Dr. Gastón Aguilera (Reproduced with permission).

Common name(s): Three-Barbeled Catfish.

Global distribution: Tropical; South America: Paraguay River basin in Paraguay.

Ecology: It is a demersal species.

Biology: This species has very distinct pits in its lower cheek and mandible which are its characteristic features. Its maxillary barbels reach up to the origin or middle of caudal fin rays. The pectoral spine is with straight, week dentation on its posterior surface. The adipose fin is short. It has a maximum length of 14 cm.

Venomous organs: The potential venom delivery system, which is the bony spine of this species, is Y shaped. Paired, large, circumferential glands are visible (Wright, 2012).

Rhamdia guatemalensis (Günther, 1864)

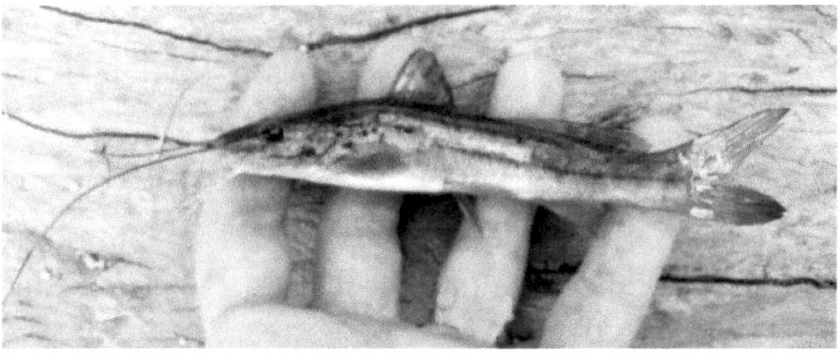

Image credit: Josh Leisen. (Reproduced with permission).

Common name(s): Pale Catfish.

Global distribution: Tropical; America: Colombia to Southern Mexico.

Ecology: It inhabits main channel of rivers and streams. It is also known to exist in pools and slow-flowing waters with plant material deposits and rocks and stones which serve as shelter for this fish especially during day time.

Biology: It has an elongated body with three pairs of barbels. Body is dusky on its sides; dark-gray dorsally, and it is white with golden tinges ventrally. The caudal and the dorsal fins are dusky and other fins are white with yellowish tinges. Some populations may have a dark stripe on their sides. The barbels are exceedingly long and extend from head to pelvic fin origin. The adipose fin is long and it originates before anus. Its caudal fin is deeply forked with wide lobes. Both the anterior and the posterior margins of the pectoral fins are serrated. It has a maximum length of 23 cm.

Venomous organs: The potential venom delivery system, which is the bony spine of this species, is Y shaped. Paired anterior and posterior glands are seen (Wright, 2012).

3.15 AIRSAC CATFISHES (HETEROPNEUSTIDAE)

Heteropneustes fossilis (Bloch, 1794)

Source: Public domain

Common name(s): Stinging Catfish.

Global distribution: Asia: Pakistan; Sri Lanka to Myanmar.

Ecology: This demersal species is commonly found in muddy rivers, flooded lakes, ditches, swamps, and marshes. It can tolerate slightly brackish water.

Biology: The body of this species is elongated and compressed. Its abdomen is more rounded. There are four pairs of barbels, that is, one maxillary pair, one nasal pair and two pairs of mandibular barbels. Its rayed fin is short with six to seven rays. Anal fin has 60–79 rays and there are 6 rays in each if its ventral fins. There are two elongate pulmonary sacs which are running backward from the gill. The body is brownish in coloration along with two yellowish, narrow longitudinal bands. Numerous black spots are seen on its flanks. The fins are often fawn. The anal fin is occasionally with dark marbling and its eye is yellow. It is an omnivorous species. It grows to a length of 30 cm.

Venomous organs: This species is considered very dangerous to humans as its pectoral-fin spine is very strongly serrated and connected to a poisonous gland (Thalgaspitiya et al., 2017).

Heteropneustes microps (Günther 1864)

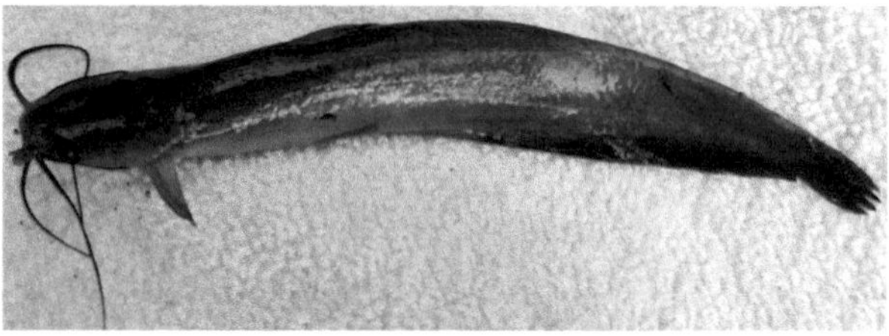

Image credit: Theo Modder Wikipedia. https://creativecommons.org/licenses/by/3.0/

Common name(s): Small Asian Stinging Catfish.

Global distribution: Tropical; Asia: Sri Lanka, India.

Ecology: Adults occur mainly in irrigation ditches and swamps. This demersal, freshwater species may enter brackish water also.

Biology: The body depth of this species equals to the head length. Its terminal mouth has well developed lips. The pectoral spine is separated from rays and is serrated along its inner edge. The anal fin is confluent with caudal without notch. It is an omnivore with nocturnal feeding habits. It attains a maximum length of 20 cm.

Venomous organs and envenomation: This species is able to deliver a stinging protein (known as poison) released from the venom glands located around its pectoral spine. In humans, this catfish venoms, which are reported to be neurotoxic and hemolytic (i.e., destruction of red blood cells), may produce local numbness, inflammation, and severe painful physical conditions of the human limb when the latter is under attack (https://www.fws.gov/fisheries/ans/erss/uncertainrisk/ERSS_Heteropneustes_microps_final_January_2018.pdf)

3.16 NORTH AMERICAN CATFISHES (ICTALURIDAE)

Ameiurus brunneus (Jordan, 1877)

Image credit: Zach Randall, Florida Museum of Natural History. (Reproduced with permission)

Common name(s): Snail Bullhead.

Global distribution: Atlantic Slope: Georgia, North Carolina, Alabama, Florida, and Virginia.

Ecology: This species is found in the flowing pools of streams and rocky riffles.

Biology: It has a flat head and a well-rounded snout. Its body is yellowish brown or olive dorsally and belly is blue white to white. A large dark blotch is seen at the base of the dorsal fin. The fins are dusky olive brown. A narrow black edge is seen in all fins excepting pectoral. The rear edge of the pectoral spine lacks saw-like teeth. The anal fin is short and rounded with about 20 rays. The first gill arch has 14–17 rakers. It has a total length of 29 cm.

Venomous organs: The potential venom delivery system, which is the bony spine of this species, is Y shaped. There is no discernible glandular cell (Wright, 2012).

Ameiurus catus (Linnaeus, 1758)

Image credit: Zach Randall, Florida Museum of Natural History. (Reproduced with permission)

Common name (s): White Catfish.

Global distribution: North America: Atlantic coastal states of the USA.

Ecology: This freshwater, brackish, demersal and potamodromous species is largely found in backwaters, open channels and in stagnant muddy pools. It may also occur in lakes and impoundments.

Biology: This species is normally gray to blue black above and white to light yellow below. It is largely distinguished by its dusky or black adipose fin, shorter anal fin base, and lower degree of forking in the caudal fin. There are two nasal, two maxillary and four chin barbels on its head.

In this naked fish, there are 21 rakers on the first gill arch. It attains a maximum length of 95 cm.

Food and feeding: It feeds on a wide variety of fishes, insects, and crustaceans.

Reproduction: In this species, reproduction occurs during the period April–July at a water temperature of about 70°F. At the time of spawning, a cavity is created using hollow logs where a gelatinous mass of eggs is deposited. Male is believed to guard the nest and incubate the eggs by splashing fresh water over such eggs.

Venomology: Its dorsal and pectoral fins bear a venomous spine on their anterior edge. The rear edge of its pectoral-fin spine has 15 saw-like teeth. Its venom has been reported to show lethal activity in mice. Further, toxic peptides obtained from the fin spine extracts of this species displayed a considerable toxic effect in the fish, largemouth bass (Wright, 2015).

Ameiurus melas (Rafinesque, 1820)

Image credit: Dr. Uland Thomas (Reproduced with permission).

Common name(s): Black Bullhead; Black Bullhead Catfish.

Global distribution: Temperate; North America; Europe: Ebro and Tagus drainages.

Ecology: This benthic species dwells in small-to-large rivers and creeks; also in ponds and impoundments with soft substrates.

Biology: It is a dorsoventrally flattened fish without scales. The body coloration is generally dark. It has a broad head and slightly humped back.

The lower lip is shorter than the upper lip. The barbels are black and are located near its mouth. The anal fin has a gray base, and the tail fin which is squared/truncated has a pale bar. Its pectoral spine is smooth. It is a nocturnal feeder and adults feed mainly on clams, snails, plant material, and fishes. It has a maximum length of 66 cm and weight of 3.6 kg.

Venomous organs and envenomation: The potential venom delivery system, which is the bony spine of this species, is Y shaped. There is no discernible glandular cell (Wright, 2012). The spines present at the base of its pectoral and dorsal fins are used as spurs to cut predators. Extracts from the stings of this species displayed lethality to certain species of fish (Birkhead, 1972).

Ameiurus natalis (Lesueur, 1819)

Source: Public domain

Common name(s): Yellow Bullhead.

Global distribution: Temperate; North America: New York to Mexico; Quebec to North Dakota.

Ecology: This demersal species inhabits small-to-large rivers; pools, ponds, creeks, backwaters, and impoundments with soft substrate.

Biology: It is a medium-sized species of the catfish family with a maximum length of 60 cm. The body coloration is olive to dull black on the back; lateral sides are yellowish; ventrum and chin are bright yellow to white. The rear edge of its caudal fin is rounded and anal fin is much larger. Its white barbels or "whiskers" are the characteristic feature of this species. It feeds on insects and bottom dwelling mollusks and crustaceans. It has a maximum weight of about 2 kg.

Venomous organs: The potential venom delivery system, which is the bony spine of this species, is Y shaped. Paired glands are seen in anterior grooves (Wright, 2012).

Ameiurus nebulosus (Lesueur, 1819)

Image credit: Kara Benson. Wikimedia Commons. https://creativecommons.org/licenses/by-sa/2.5/

Common name(s): Brown Bullhead; Mud Cat.

Global distribution: Subtropical; North America: Canada and USA.

Ecology: This demersal species inhabits creeks and small-to-large rivers; impoundments, lakes, pools, and ponds at a depth range of 0–12 m. It may occasionally enter brackish waters.

Biology: This species has a single adipose fin, two dorsals, and a single anal with 24 rays. Tail is only slightly notched and its dorsal and ventral lobes are angling inward. Its body is without scales and its coloration is darker brown green dorsally. Its belly is off-white or cream. The entire surface of the fish may have brown black speckles. The barbels are seen on the pelvic spine and around its mouth. The chin barbels are black to thin brown. It is a nocturnal feeder consuming mainly mollusks, insects, leeches, crayfish, and worms. It may also feed on algae, plant material and fishes and their eggs. It may attain a maximum length of 55 cm and weight of 2.7 kg.

Venomous organs: The potential venom delivery system, which is the bony spine of this species, is Y shaped. Paired glands are seen in anterior grooves (Wright, 2012).

Ameiurus platycephalus (Girard, 1859)

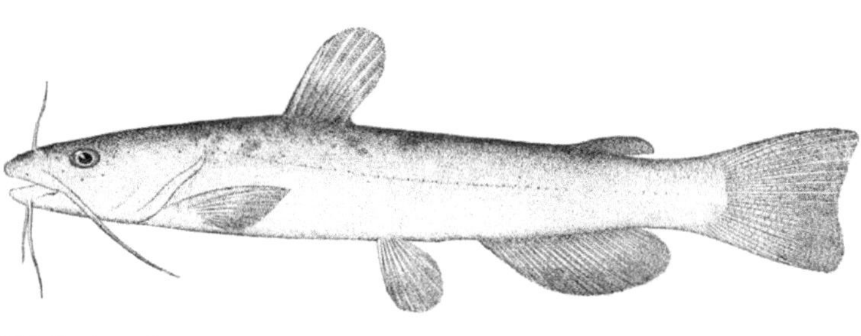

Source: Public domain

Common name(s): Flat Bullhead.

Global distribution: Temperate; North America: USA.

Ecology: This demersal species occurs in small-to-large rivers with rock-bottomed pools. It is also found in lakes, impoundments, and ponds.

Biology: As its name suggests, it has a flat head. The rear edge of the pectoral spine lacks saw-like teeth, and the snout has a more-or-less straight profile. Its body is golden yellow to dark brown dorsally and belly is dusky cream to white. The fins are dusky black and lateral sides are with dark brown spots. The base of the dorsal fin base bears a large dark blotch. The adipose, dorsal, anal, and the caudal fins are with black edge. The anal fin is short, with 21–24 rays and its first gill arch has 11–13 rakers. It feeds mainly on aquatic insects, small fishes, and snails and it reaches a maximum length of 29 cm.

Venomous organs: The potential venom delivery system, which is the bony spine of this species, is Y shaped. Paired glands are seen in anterior grooves (Wright, 2012).

Ameiurus serracanthus (Yerger and Relyea, 1968)

Source: Public domain

Common name(s): Spotted Bullhead.

Global distribution: Subtropical; North America: Florida, Georgia, and Alabama in the USA.

Ecology: This demersal species occurs in rock- or sand-bottomed pools of small-to-medium fast-flowing rivers and impoundments.

Biology: It is a small strikingly marked catfish with a black blotch in the base of its dorsal fin. Eyes are large. It has its characteristic light-colored spots on the dark body. Body and fins are suffused with yellow. The barbels are dusky to dark. It has a strongly serrated pectoral spine which has 6–20 large tooth-like projections on the posterior margin. All the fins are edged in black, and e caudal fin is fairly emarginate. It has a maximum length of 28 cm.

Venomous organs: The potential venom delivery system, which is the bony spine of this species, is Y shaped. The glands are in four distinct pairs that are surrounding the spine (Wright, 2012).

Ictalurus furcatus (Valenciennes, 1840)

Image credit: Thomsonmg2000, Wikimedia Commons. https://creativecommons.org/publicdomain/zero/1.0/deed.en

Common name (s): Blue Catfish.

Global distribution: Subtropical; North America: USA: South Dakota, Nebraska, Alabama, Florida, Texas, and New Mexico.

Ecology: This demersal and brackish water species dwells in the backwaters associated with muddy, sandy, or gravelly rivers; main channels and impoundments with deep water.

Biology: It is distinguished by its congener, namely, *Ictalurus punctatus* by its anal fin which has a straight edge with a maximum of 36 rays. This heavy bodied fish has blueish gray coloration and it has a dorsal hump. It also has a well-forked tail fin, a protruding upper jaw, and barbels. Its diet includes clams, fishes, and small aquatic invertebrates. It may take shelter in deep waters during the day and move into fast-flowing waters during night to feed. It reaches a maximum length of 165 cm and weight of 68 kg.

Venomous organs: The potential venom delivery system, which is the bony spine of this species, is Y shaped. Venom glands are paired and are in anterior grooves (Wright, 2012).

Ictalurus pricei (Rutter, 1896)

Image is not available.

Common name(s): Yaqui Catfish.

Global distribution: Subtropical; North America: Mexico and USA.

Ecology: Freshwater; this demersal species occurs in small-to-medium rivers sandy or rocky bottom.

Biology: The coloration of the head, dorsum, and sides are clear brown. The fins are gray in coloration. Its caudal fin has a black edge and tips of its lobes are pointed. It has a maximum length of 57 cm.

Venomous organs: The potential venom delivery system, which is the bony spine of this species, is Y shaped. Anterior groove of the spine is visible and glandular cells are, however, absent. Glandular tissue is visible posteriorly (Wright, 2012).

Ictalurus punctatus (Rafinesque 1818)

Source: Public domain

Common name(s): Channel Catfish.

Global distribution: North America: Missouri–Mississippi river basins, St. Lawrence-Great Lakes and Hudson Bay; Northern Mexico; Eastern New Mexico.

Ecology: It is a freshwater, demersal species inhabiting depths of 0–15.

Biology: It is a bilaterally symmetrical fish without scales. It is distinguished by its deeply forked caudal fin. The body of the fish is spotted; the coloration of the back is black and the ventrum is whitish. However, the body coloration of this fish may change from blue to black. The upper jaw and the lower jaw bear two and four barbels, respectively. Males are darker in body coloration and are with large heads. It grows to a maximum length of 127 cm.

Venomous organs: The dorsal and pectoral fins of this venomous species have hard spines, which are serrated. Its venom gland is with a shaft (Egge and Simons, 2011).

Noturus albater (Taylor, 1969)

Image credit: Dr. Larry M. Page (Reproduced with permission).

Common name(s): Ozark Madtom.

Global distribution: Temperate; North America: Missouri and Arkansas, USA.

Ecology: This demersal species occurs in small-to-medium rivers and pools of creeks.

Biology: It has a stout body and a short head. A dark bar is present on the base of the caudal fin. The caudal fin is usually straight or slightly rounded. Also, the caudal fin has irregular dark bands and a creamy white area is seen on its upper margin. The pectoral spine has smaller teeth on its front edge and 6–10 larger teeth on its rear edge. Its body is yellowish brown to light brown dorsally and belly is pale yellow to cream white. Dark spots or blotches and four dusky saddles are seen. The fins are yellow or white with weak blotches. Its diet largely consists of aquatic insects especially mayflies. It has a total length of 12 cm.

Venomous organs: The potential venom delivery system, which is the bony spine of this species, is Y shaped. Venom glands are paired and are present in anterior grooves (Wright, 2012).

Noturus baileyi (Taylor, 1969)

Image is not available.

Common name(s): Smoky Madtom.

Global distribution: Temperate; North America: Tennessee, USA.

Ecology: This demersal species occurs in flowing pools of creeks runs and rocky riffles. This species relies largely on clean water and undisturbed rocks to complete its life cycle.

Biology: It is olive brown on dorsally and belly is white to yellow. Four saddles are found to line the back of this fish. It is a small sized fish with a total length of 7 cm. This species was considered extinct until the early 1980s and was considered "federally endangered." It is now rediscovered (https://www.wbtv.com/2019/10/16/venomous-fish-with-painful-sting-was-presumed-extinct-smoky-mountains-its-back).

Venomous organs: It has venomous spines that can deliver a painful sting to humans. Its sting is generally considered equivalent to a bee sting (https://www.wbtv.com/2019/10/16/venomous-fish-with-painful-sting-was-presumed-extinct-smoky-mountains-its-back).

Noturus elegans (Taylor, 1969)

Image credit: Drs. David and Lynn Eisenhour (Reproduced with permission).

Common name(s): Elegant Madtom.

Global distribution: Temperate; North America: Kentucky and Tennessee, USA.

Ecology: This benthopelagic species inhabits rocky riffles clear creeks and small rivers.

Biology: It has a long, slender body which is yellow-gray to brown above three or four dark saddles are seen on the body. A brown blotch is present on the base of the dorsal fin and it is found extending up to its front edge. The lower half of the adipose fin bears a dark blotch. The caudal fin has one or two dark bands on its edge and one in the middle part of the fin. The adipose fin has its origin is over the anal fin, and it is well connected to the caudal fin. The pectoral spine is short with small teeth along the front and five to nine large teeth on the rear edge. The caudal fin is slightly rounded. This fish has a maximum total length of 7 cm.

Venomous organs: The potential venom delivery system, which is the bony spine of this species, is Y shaped. Venom glands are paired and are present in anterior grooves (Wright, 2012).

Noturus eleutherus (Jordan, 1877)

Image credit: Renn Tumlison (Reproduced with permission).

Common name(s): Mountain Madtom.

Global distribution: Temperate; North America: from Pennsylvania to Illinois; Tennessee and Georgia; and Missouri and Arkansas, USA.

Ecology: This demersal species inhabits clean rocky riffles and small-to-large rivers in dense vegetation attached to the bottom materials such as stones, rubble, gravel, and sand. It seems to be very sensitive to siltation and other pollutants.

Biology: This species is a small, slender catfish. Its fins and sides are with yellow spots. Three brownish dorsal saddles may be seen. Its adipose fin is connected with the tail fin, but it appears to be separate. There are slightly curved pectoral spines which possess small anterior serrae and large and sharp posterior serrae. It is a nocturnal feeder feeding on larvas of aquatic insects particularly mayflies, caddisflies, and stoneflies. It attains a maximum length of 13 cm.

Venomous organs: The potential venom delivery system, which is the bony spine of this species, is Y shaped. Spine is with well-developed anterior groove. It lacks glandular cells and glandular tissue is visible posteriorly (Wright, 2012). Its toxic sting is associated with its pectoral and dorsal spines.

Noturus exilis (Nelson, 1876)

Image credit: Marshal Hedin. Wikipedia. https://creativecommons.org/licenses/by-sa/2.0/

Common name(s): Slender Madtom.

Global distribution: Temperate; North America: USA.

Ecology: This demersal fish inhabits small rivers, rocky riffles, and flowing pools of clear creeks.

Biology: This long and slender catfish has a scaleless body and it has eight distinct barbels around its mouth. Fishes of cool clear habitats are slenderer than those from warmer waters. The head is flattened and mouth is terminal with equal jaws. The body coloration is continuous gray to dusky yellow dorsally. Its belly is light yellow to whitish. The dorsal, anal, and caudal fins are generally pale with a partial or entire black margin. Other fins are also pale or yellowish. The pectoral fins possess six prominent serrae (teeth) along their rear edge. It feeds on aquatic insects and small crustaceans such as isopods, amphipods, and copepods. It has a maximum length of 15 cm.

Venomous organs and envenomation: The serrae present on its pectoral spines may give most painful stings. These serrae may cause swelling which would last for many days (https://animaldiversity.org/accounts/Noturus_exilis/). The extracts of the stings of this species have also been reported to be lethal to Gambusia fish (Birkhead, 1972).

Noturus fasciatus (Burr et al., 2005)

Source: Public domain

Common name(s): Saddled Madtom.

Global distribution: Subtropical; North America: USA.

Ecology: It occurs in rocky areas with cobble, gravel or rubble at a mean depth of 0.2 m. Riffles are normally occupied during daylight hours by these fishes which bury themselves within the cavities of the gravel bed.

Biology: It has a slender body shape and there are three yellow eclipses on its nape. The dorsal side of the fish is with brown saddles which are not continuous. The adipose fin blotch is extending to middle of fin. The ventral edge of caudal peduncle bears a small, dark saddle. The adipose fin is weakly connected to caudal fin. A crescent-shaped dark blotch is seen at caudal fin base. The anal fin is more elongated with 18 rays. It attains a maximum length of 8.5 cm. It is an endangered species (https://www.fishbase.se/summary/Noturus-fasciatus.html).

Venomous organs: The venom delivery structures of this species include spine, venom gland, and sting. The spine is serrated; venom gland is with shaft and serration; and stings are three in number (Egge and Simons, 2011).

Noturus flavater (Taylor, 1969)

Image credit: North American Native Fishes (NANFA) Gallery. NANFA.org

Common name(s): Checkered Madtom.

Global distribution: Temperate; North America: Missouri and Arkansas, USA.

Ecology: This demersal species occurs in pools and backwaters of clear small-to-medium rivers. It is usually found among leaves and woody debris.

Biology: It has a stout body which is yellow with four clear saddle-like black stripes on top. Its belly is white to yellow. The dorsal fin on the back possesses a black blotch on the top. The caudal fin which is usually straight or slightly rounded has a straight black line at its end and base. It is attaining a maximum length of 18 cm and is said to be the second largest species of madtom.

Venomous organs: The venom delivery structures of this species include spine, venom gland, and sting. The spine is serrated; venom gland is with shaft and serration; and stings are three in number (Egge and Simons, 2011).

Noturus flavipinnis (Taylor, 1969)

Image is not available.

Common name(s): Yellowfin Madtom.

Global distribution: Temperate; North America: Virginia, Tennessee, and Georgia in USA.

Ecology: This demersal species inhabits in clear creeks, small rivers, backwaters, and pools.

Biology: Its dorsum and fins have a yellowish tinge. The dorsal fin origin and the base of the adipose fin and dorsal fin bear dark blotches. While caudal fin base has a crescent-shaped crossing dark bar, dorsal fin has a dark medial stripe. The pectoral spine is with well-developed anterior and posterior serrae. Two internasal pores are found. Feeding in this species takes place mainly during night and its diet consists mostly of aquatic insect larvas and crayfish. This species may exhibit some preferences in It is also an opportunistic feeder attaining a maximum length of 15 cm. It is a vulnerable species (https://www.fishbase.se/summary/3032).

Venomous organs: The venom delivery structures of this species include spine, venom gland, and sting. The spine is serrated; venom gland is with shaft and serration; and stings are three in number (Egge and Simons, 2011).

Noturus flavus (Rafinesque, 1818)

Source: Public domain

Common name(s): Stonecat.

Global distribution: Temperate; North America: Canada and USA.

Ecology: This demersal species inhabits rubble and boulder riffles; small and large rivers; creek runs; and gravel shoals of lakes.

Biology: It is a small, slender, flat-headed catfish. Its skin is thick and is yellowish brown in color. Sides of the head are yellow and belly is whitish. Its keel-like adipose fin is almost extending up to the caudal fin. The upper jaw protrudes beyond the lower jaw. These fishes occur in two forms. While the fishes of in Tennessee bear two faint bars on the nape, fishes of other localities are with a patch. However, both the forms are provided with a white spot on their caudal and dorsal fins. The rear of the pectoral spine may or may not have a few weak teeth. The adult fishes feed on mayfly larvas and crayfish. It has a maximum length of 31 cm.

Venomous organs: In this species, poison glands are seen at the bottom of its pectoral spine. The potential venom delivery system, which is the bony spine of this species, is Y shaped. Venom glands are paired in poorly-developed anterior groove (Wright, 2012). The extracts from the stings of this species have been reported to be quite virulent (Birkhead, 1972).

Noturus funebris (Gilbert and Swain, 1891)

Image credit: Bernie Kuhajda (Reproduced with permission).

Common name(s): Black Madtom.

Global distribution: North America; Florida, Mississippi, and Louisiana in USA.

Ecology: This demersal species lives in gravel and sandy substrates in creeks to small rivers with aquatic vegetation.

Biology: It has a stout body which is black or blue black above and lighter below. Although black spots are seen all over the body, they are distinct. These dots are very distinct on its head and belly. The anal fin is long and it reaches the caudal fin. The upper jaw projects beyond the lower jaw. A few small teeth are seen on the rear edge of its pectoral spine. The caudal fin is straight or slightly rounded. It attains a maximum length of 15 cm.

Venomous organs: The venom delivery structures of this species include spine, venom gland, and sting. The spine is smooth; venom gland is with shaft and there is a single sting (Egge and Simons, 2011).

Noturus furiosus (Jordan and Meek, 1889)

Image credit: Wikimedia Commons.

Common name(s): Carolina Madtom.

Global distribution: Temperate; North America: North Carolina, USA.

Ecology: This demersal species inhabits small-to-medium river runs and sand, gravel and detritus-bottomed riffles.

Biology: It has a short and bulky body with a definite coloration. There are three dark saddles in its back and these unite with a black stripe present on its side. The adipose fin has a dark blotch which does not reach the fin's edge. While the saddles are yellowish, tan coloration prevails in the rest of the fish. The edge and center of its tail fin has moon -shaped brown bands. This nocturnal feeder feeds on benthic animals like caddisflies, dragonflies, larval midges, beetle larvas and mayflies. This fish attains a total length of 12 cm.

Venomous organs: The potential venom delivery system, which is the bony spine of this species, is Y shaped. Large hemispherical venom glands are seen (Wright, 2012).

Noturus gilberti (Jordan and Evermann, 1889)

Source: Public domain

Common name(s): Orangefin Madtom.

Global distribution: North America: Virginia and North Carolina, USA.

Ecology: This freshwater, demersal species dwells small river runs and rocky riffles.

Biology: It has a slender body and a flat head which has a protruding upper jaw. The body coloration is brown dorsally and light yellow ventrally. The fins are dark colored with light yellow or white edges. The caudal fin is dusky to dark and its upper edge has a clear-cut white triangle. The anal fin is short with a maximum of 16 rays. The caudal fin edge may be straight or slightly rounded. It grows to a maximum length of 10 cm.

Food and feeding: It is an opportunistic insectivore feeding largely on immature stages of benthic aquatic insects. It is often found under small rocks and stones of shallow stream riffles in search of its food.

Reproduction: It is a nester and egg guarder.

Venomous organs: The pectoral and dorsal spines of this venomous species are short, and the pectoral spine has large teeth on its rear edge.

Noturus gladiator (Thomas and Burr, 2004)

Image is not available.

Common name(s): Piebald Madtom.

Global distribution: Subtropical: North America: Tennessee and Mississippi, USA.

Ecology: This demersal species lives in the clean sand or clay substrata of small-to-medium-sized mainstream rivers and in their major tributaries with a depth range, 0–1 m.

Biology: It has dark brown to black dorsal saddles against its pale yellow to tan areas on body. The dorsal fin is with dark, broad submarginal and basal bands. The anal fin has dark central blotch which is connected to intense dark pigment on sides of caudal peduncle. A dark, crescent-shaped mid-caudal band is also seen. The pectoral spine has strongly developed anterior dentations and some are bifurcate. In its pectoral spine, posterior serrae, and posterior process of cleithrum are very large. The pectoral fin is long and sickle shaped, and when it is depressed, it may reach beyond pelvic fin origin. It attains a maximum length of 11 cm.

Venomous organs: The venom delivery structures of this species include spine, venom gland, and sting. The spine is serrated. The venom gland is with shaft and serration; and stings are three in number (Egge and Simons, 2011).

Noturus gyrinus (Mitchill, 1817)

Source: Public domain

Common name(s): Tadpole Madtom.

Global distribution: Temperate; North America: from New Hampshire to Texas and Gulf of Mexico, USA; Quebec to Saskatchewan in Canada.

Ecology: This demersal species inhabits rock- or mud-bottomed pools and lakes; and backwaters of rivers and creeks.

Biology: Its body of this species is heavy, round, and potbellied anteriorly and is strongly compressed posteriorly. It has a dark brown dorsum and white belly. The pelvic and pectoral fins are heavily covered in melanophores and dorsal and anal fins however possess fewer melanophores. The bands of melanophores are also seen on both the upper and the lower jaws and they look like dark streak along its side. The barbels of chin may be white or lightly covered with melanophores. The nasal and maxillary barbels are dark and mandibular barbels are white. The adipose fin is entirely connected to its large and round caudal fin, and the pectoral fin is not serrated. It possesses 67 gill rakers. The anal fin is of moderate length but it gets reduced with increasing body length. Its terminal mouth is with numerous small and sharp cardiform teeth. Its diet includes isopods, amphipods, midge, mayfly, and caddis larvas. It reaches a maximum length of 13 cm.

Venomous organs: It is one of the more venomous catfishes with toxic (neurotoxic and haemotoxic) sheath covering dorsal and pectoral spines. Threatened fish will arch back and erect its sharp spines. The effects of stings of these spines on humans are comparable to that of bee stings (http://txstate.fishesoftexas.org/noturus%20gyrinus.htm).

Noturus hildebrandi (Bailey and Taylor, 1950)

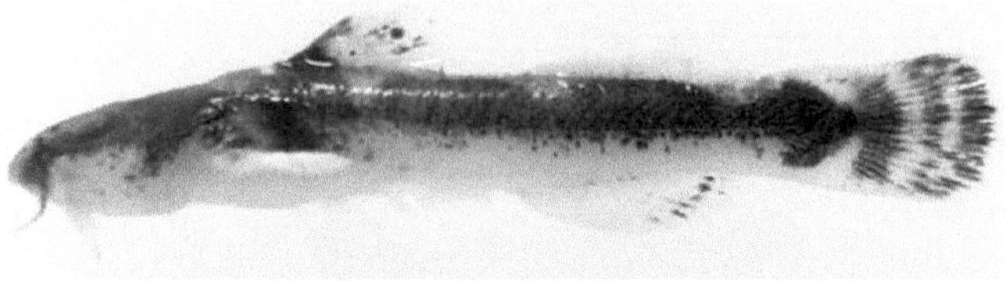

Image credit: Lawrence M. Page (Reproduced with permission).

Common name(s): Least Madtom.

Global distribution: Subtropical: North America: tributaries of Mississippi River, USA.

Ecology: This demersal species occurs in small rivers and creeks; and sand or rock riffles. It is often seen near debris.

Biology: The body of this species is slender with a flat head. The adipose fin is white or clear; pectoral fin is short with four to five large saw-like teeth on its rear border. The caudal fin's rear border is straight or with slightly rounded corners the adipose and caudal fins are united with a tiny notch between. There are two subspecies in this species. While its subspecies *Noturus hildebrandi* latus has a shorter head, *Noturus hildebrandi* possesses longer head. It feeds mainly on caddis larvas, mayfly, and stonefly nymphs. It attains a maximum length of 7 cm.

Venomous organs: The venom delivery structures of this species include spine, venom gland, and sting. The spine is serrated; venom gland is with shaft; and stings are two in number (Egge and Simons, 2011).

Noturus insignis (Richardson, 1836).

Source: Public domain

Common name(s): Margined Madtom.

Global distribution: Temperate; North America: New York to Altamaha; Georgia, Virginia, and North Carolina, USA.

Ecology: This species inhabits rocky riffles, rivers and creeks. It has been reported to seek cover during the day and forage actively during the night hours.

Biology: The body of this species is yellow to slate gray dorsally and belly is white or light. The fins are yellow or light gray and the median fins other than the adipose fin are black bordered. While the fin borders of the fish of turbid waters are dusky, it is dirty black in clear streams. The upper jaw projects beyond the lower jaw. The rear edge of pectoral spine has large teeth. The caudal fin border is fairly rounded or straight. It attains a total length of 15 cm.

Venomous organs: The potential venom delivery system, which is the bony spine of this species, is Y shaped. The venom glands are paired and are present in the anterior grooves of the spine (Wright, 2012).

Noturus lachneri (Taylor, 1969)

Image is not available.

Common name(s): Ouachita Madtom.

Global distribution: North America: Arkansas in USA.

Ecology: It inhabits rocky pools, creeks, and small rivers.

Biology: It has a slenderer body with a short and flat head. Its body is dark gray or brown above and white below. Its fins may have dark borders. It may attain a maximum length of 10 cm.

Venomous organs: The venom delivery structures of this species include spine, venom gland, and sting. The spine is smooth; venom gland is with shaft and there is a single sting (Egge and Simons, 2011).

Noturus leptacanthus (Jordan, 1877)

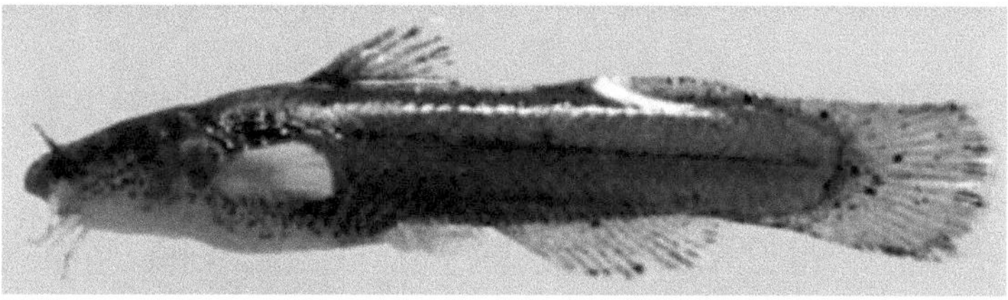

Image credit: Lawrence M. Page (Reproduced with permission).

Common name(s): Speckled Madtom.

Global distribution: Subtropical; North America: South Carolina, Louisiana, and Florida, USA.

Ecology: This demersal species dwells in the riffles of rivers and creeks. It is frequently seen with debris or plants.

Biology: It has a slender body, coloration of which is reddish or yellowish brown dorsally with several distinct black specks. Its belly is light yellow to white without black specks. The dorsal, anal, caudal, and adipose fins are dark and their borders are with light coloration. A dark blotch is seen at the base of the pectoral fin. The upper jaw projects beyond the lower jaw. Its short pectoral spine is without teeth. The caudal fin has a straight or rounded edge. It feeds largely on midge larvas. It has a maximum length of 9.4 cm.

Venomous organs: The potential venom delivery system, which is the bony spine of this species, is Y shaped. Venom glands are paired and are present in the anterior grooves of the spine (Wright, 2012).

Noturus maydeni (Egge, 2006)

Image credit: North American Native Fishes (NANFA) Gallery. NANFA.org

Common name(s): Black River Madtom; Eastern Ozark madtom.

Global distribution: Temperate; North America: Arkansas and Missouri, USA.

Ecology: This demersal species inhabits moderate-to-large high gradient streams with gravel, cobble, or large stone substrate. It may also be seen in shallower pools.

Biology: It is a small, slender species with a scaleless body and rounded head. Its subcutaneous eyes are directed dorsally. The adipose fin is adnate. The pectoral fin is short with many, small serrations on its posterior edge. The dorsal fin is with one spine. The upper jaw protrudes well. The body coloration is gray brown with pale yellow background and belly. The body and fins are mottled. Three dorsal saddles cross the back. The caudal fin base is densely pigmented. The dorsal fin is also pigmented. Its caudal fin has a dark margin posteriorly. It is an insectivorous species. It attains a maximum length of 8 cm.

Venomous organs: The venom delivery structures of this species include spine, venom gland, and sting. The spine is serrated; venom gland is with shaft and serration; and stings are three in number (Egge and Simons, 2011).

Noturus miurus (Jordan, 1877)

Source: Public domain

Common name(s): Brindled Madtom.

Global distribution: Subtropical; North America: Ontario, Canada; and New York, Mississippi and Louisiana, USA.

Ecology: This demersal species inhabits riffles, pools, creeks, small rivers, and lakes.

Biology: It is a scaleless species with eight whisker-like barbels around its mouth. Its head is rounded and flattened with large eyes and a wide mouth. The dorsal and pectoral fins possess venomous spines. While small, saw-like notches (serrae) are seen on the front portion of the pectoral fins, posterior portion of the pectoral fins is provided with large teeth. The upper jaw protrudes slightly. The body coloration is typically a yellow-brown to gray. There are –three to five dark saddles which are extending over its back. The lateral line and the spaces between the saddles are with a light pigmentation. The body coloration is becoming lighter toward the lateral surface. The belly is distinctly white. Maxillary and nasal are dark brown. The pelvic and pectoral fins are with scattered pigmentation. A black spot is seen along the anterior of the dorsal fin. The anal and adipose fins have a series of small blotches which are extending to the base of the fin. The caudal fin is wide, dark, and rounded. The adipose fin may be united with the caudal fin or they may be separated by a small notch. It feeds on larvas of midges and blackflies; mayflies; and hydropsychid caddisflies. It attains a maximum length of 13 cm.

Venomous organs: The venom delivery structures of this species include spine, venom gland, and sting. The spine is serrated; venom gland is with shaft and serration; and stings are three in number (Egge and Simons, 2011).

Noturus munitus (Suttkus and Taylor, 1965)

Source: Public domain

Common name(s): Frecklebelly Madtom.

Global distribution: Temperate; North America: Tennessee, Georgia, Alabama, Mississippi; and Louisiana, USA.

Ecology: This demersal species inhabits rocky riffles of small-to-large rivers. These fishes are mostly found in or near aquatic vegetation.

Biology: It has a stout body which is yellow to dark brown with dark mottling dorsally four dark saddles are also present on the body. In front of the dorsal fin, there are spots which are light and large. Its belly has dark specks. The fins are mottled or blotched and dorsal fin has a broad dark band nearer to its clear edge. The caudal fin has two dark crescent-shaped bands. The adipose fin has a dark brown or black band which is extending to the edge. The pectoral spine is large with about 10 large teeth on the rear edge. The large teeth are seen on the front edge. Caudal fin is straight or slightly rounded and is free from the adipose fin. It feeds mostly on hydropsychid caddisfly larvas, ephemerellid mayfly nymphs, blackfly, and midge larvas. It attains a total length of 10 cm.

Venomous organs: The potential venom delivery system, which is the bony spine of this species, is Y shaped. Venom glands are paired and are present in the anterior grooves of the spine (Wright, 2012).

Noturus nocturnus (Jordan and Gilbert, 1886)

Image credit: Dr. Garold W. Sneegas (Reproduced with permission).

Common name(s): Freckled Madtom.

Global distribution: Subtropical; North America: Illinois to Louisiana; Kentucky to Kansas and Oklahoma; and Alabama to Texas, USA.

Ecology: This demersal fish is often found in creeks to large rivers.

Biology: It is dark brown or gray dorsally with very small dark colored dots all over the fins and body. Its belly is light yellow or white without dark dots or specks. The anal fin has a black edge; and other fins are dark at the base. The posterior border of the pectoral spine is with re two to three small teeth. The rear edge of caudal fin is straight or slightly rounded. It feeds largely on aquatic insects such as mayflies, caddisflies, midges, and blackflies. It attains a maximum length of 15 cm.

Venomous organs and envenomation: The potential venom delivery system, which is the bony spine of this species, is Y shaped. Venom glands are paired and are present in the anterior grooves of the spine (Wright, 2012). The extracts from the stings of this species have been reported to be quite virulent (Birkhead, 1972).

Noturus phaeus (Taylor, 1969)

Image credit: Bernie Kuhajda (Reproduced with permission).

Common name(s): Brown Madtom.

Global distribution: Temperate; North America: USA.

Ecology: This demersal species inhabits sand–gravel riffles, springs, creeks, and small rivers.

Biology: It has a stout body which is light to dark brown above and lighter below. A pattern of tiny brown dots is seen all over the body and these dots are most conspicuous on the light underside of the head and belly. Its anal fin is long and it reaches the caudal fin. The upper jaw projects beyond the lower jaw. The pectoral spine is with six large teeth on its rear edge. The caudal fin is straight or slightly rounded. It attains a maximum length of 15 cm.

Venomous organs: The venom delivery structures of this species include spine, venom gland, and sting. The spine is serrated; venom gland is with shaft; and stings are two in number (Egge and Simons, 2011).

Noturus placidus (Taylor, 1969)

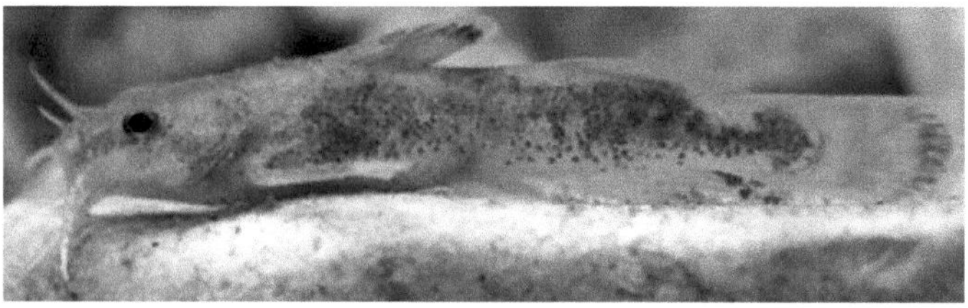

Source: Public domain

Common name(s): Neosho Madtom.

Global distribution: Temperate; North America: Arkansas, Missouri, Kansas, and Oklahoma, USA.

Ecology: This demersal species occurs in gravel riffles and runs of small-to-medium rivers.

Biology: It has a stout body and a deep caudal peduncle. The body coloration is dull pink with brown spots. Saddles and blotches are also seen. The belly is white to yellow. The rays along the lower caudal fin are white. A weak pair of light spots may or may not be present in front of the dorsal fin. The pectoral spine is large and it has 6–10 large teeth on the rear edge and small teeth on the front edge. The adipose fin is connected to the caudal fin. The lower half of the adipose fin has a dusky band. The caudal fin is straight or slightly rounded and is with a dark moon-shaped band in the center and another band in its white border. It eats aquatic insects such as mayfly larvas especially before sunset. It attains a maximum length of 8.7 cm.

Venomous organs and envenomation: The potential venom delivery system, which is the bony spine of this species, is Y shaped. Venom glands are paired and are present in the anterior grooves of the spine (Wright, 2012). The extracts from the stings of this species have been reported to be quite virulent (Birkhead, 1972).

Noturus stigmosus (Taylor, 1969)

Image credit: Lawrence M. Page (Reproduced with permission).

Common name(s): Northern Madtom.

Global distribution: Temperate; North America: Pennsylvania, Michigan, Kentucky to Ohio, Tennessee and Mississippi, USA.

Ecology: This demersal species inhabits small-to-large rivers with moderate current.

Biology: The body of this species is more dark mottling. The front and hind margins of the curved spines of its pectoral fins are well toothed. The middle of the tail fin has a dark bar. The adipose fin is fairly large. It forages at night and is feeding largely on aquatic insect larva. It has a maximum length of 13 cm.

Venomous organs: The potential venom delivery system, which is the bony spine of this species, is Y shaped. Venom glands are paired and are present in the anterior grooves of the spine (Wright, 2012).

Noturus taylori (Douglas, 1972)

Image credit: Dr. Bruce H. Bauer (Reproduced with permission).

Common name(s): Caddo Madtom.

Global distribution: Temperate; North America: Southwestern Arkansas, USA.

Ecology: This demersal species inhabits rocky riffles pools and rivers.

Biology: It has a long, slender body which is yellow to dark brown above with four dark brown saddles. The front half of the border of the dorsal fin bears a large black blotch. Adults may possess dark bands near the clear edge of the anal and pectoral fins. The caudal fin has two to three faint brown crescent-shaped bands, that is, one in the middle and one to two near the clear edge of the fin. The pectoral spine is short with a maximum of nine large teeth on its rear edge and the front border bears many small teeth. The caudal fin is straight or rounded and it joins the adipose fin with a small notch between the two fins. The adipose fin has a brown band which is extending into its upper half. It attains a maximum length of 7.7. cm. It is an endangered species (https://www.fishbase.se/summary/Noturus-taylori.html).

Venomous organs: The venom delivery structures of this species include spine, venom gland, and sting. The spine is serrated; venom gland is with shaft and serration; and stings are three in number (Egge and Simons, 2011).

Pylodictis olivaris (Rafinesque, 1818)

Source: Public domain

Common name(s): Flathead Catfish.

Global distribution: Subtropical; North America: Pennsylvania, North Dakota, Louisiana, Georgia and Alabama in USA; Mexico.

Ecology: This demersal species inhabits pools with logs and debris in rivers; and impoundments and lakes. It exists in turbid and cloudy water with slow currents.

Biology: This species has a naked skin, flattened head, and barbels around its mouth. While the dorsal fin bears one single long spine, the pectoral fin has one such spine on its either side.

Dorsum is pale yellow to light brown on its back and sides, and is highly mottled with black and/or brown. The belly is pale yellow or cream colored. It has a projecting lower jaw. The tail fin is short and is slightly notched and not deeply forked. It consumes mainly crayfish, clams and fishes and attains a maximum length of 155 cm and weight of about 50 kg.

Venomous organs: The potential venom delivery system, which is the bony spine of this species, is Y shaped and no venom glands are evident in the spine (Wright, 2012).

3.17 SQUEAKERS (MOCHOKIDAE)

Chiloglanis productus (Ng and Bailey, 2006)

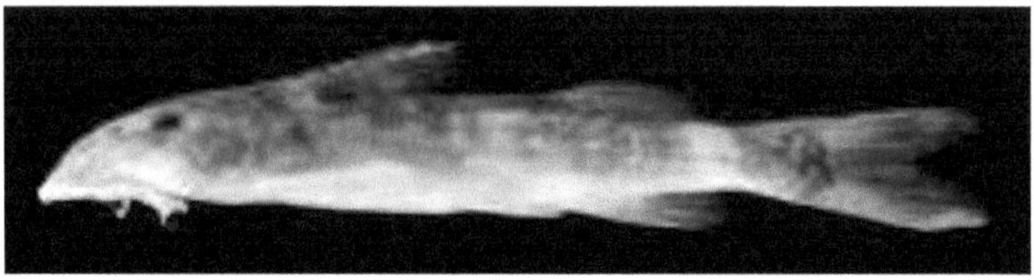

Image credit: Ray Schmidt. https://creativecommons.org/licenses/by-nc/4.0/

Common name(s): Lunzua Suckermouth Catfish.

Global distribution: Tropical; Africa: Lake Tanganyika.

Ecology: It is a demersal species.

Biology: The body of this species is cylindrical and is slightly depressed. While its predorsal profile is slightly convex, its postdorsal body is gently sloping ventrally. The preanal profile is horizontal. The skin is smooth and its lateral line is complete and mid-lateral. Its head is depressed and broad. The gill openings are narrow and are extending from ventral to base of pectoral spine. The eyes are ovoid and are located completely in dorsal half of head. Its oral sucker is round and moderately large with numerous papillae of uniform size. The barbels are well developed; maxillary barbels extend beyond hind edge of oral sucker; and auxiliary mandibular barbels are located on each side of the lower lip. The oral teeth are present in rows. The dorsal fin is with convex margin and its spine is short, stout, and laterally flattened. In adult fishes, very weak and proximally directed notches are present on distal portion of anterior edge of spine. The adipose fin is moderate and is extending for about half of the distance of postdorsal fin. The caudal fin is sexually dimorphic; males are with broad caudal fin; and in females, it is emarginate. The anal-fin base which is located below adipose fin, and also exhibits sexual dimorphism in size as males are with longer anal-fin rays. In alcohol-preserved specimens, the dorsal and lateral head and body are purplish gray; and belly and oral disc are brownish. The dorsal midline of body has two ovoid light brown patches. The dorsal margin of adipose fin is light brown. The dorsal- and anal-fin rays are purplish brown; pelvic fins are hyaline; and pectoral fins and their rays are purplish brown. The caudal-fin rays are also purplish brown, with two ovoid weak brown spots arranged in the upper and the lower bases of the caudal-fin lobes. It has a maximum length of about 5 cm.

Venomous organs: The potential venom delivery system, which is the bony spine of this species, is Y shaped. Paired anterior and posterior venom glands are seen (Wright, 2012).

Euchilichthys astatodon (Pellegrin, 1928)

Image credit: Benutzer:Haps, Wikimedia.

Common name(s): None.

Global distribution: Tropical; Africa: Congo and Angola.

Ecology: It is a demersal species.

Biology: It has a maximum length of 18.5 cm.

Venomous organs: The potential venom delivery system, which is the bony spine of this species, is Y shaped. There are no discernible glandular cells (Wright, 2012).

Microsynodontis batesii (Boulenger, 1903)

Source: Public domain

Common name(s): Dwarf Synodontis Catfish; Squeaker or Upside-Down Catfish.

Global distribution: Tropical; Africa: Gabon and Cameroon.

Ecology: It is a demersal species.

Biology: In this species, its dorsal and lateral surfaces of head are provided with short tubercles. The dorsal spine is gently curved. The adipose fin is very long and caudal peduncle is slender. Anterior edge of pectoral spine is with antrorse serrations. The coloration of dorsal surface is gray to olive brown or black. Five yellowish to cream irregular vertical stripes are seen extending from its dorsum toward the belly. It has a maximum length of 8.5 cm.

Venomous organs: The potential venom delivery system, which is the bony spine of this species, is Y shaped. No venom glands are visible in the spine (Wright, 2012).

Mochokiella paynei (Howes, 1980)

Image credit: Ingo Seidel (Reproduced with permission).

Common name(s): Payne's Catfish.

Global distribution: Tropical; Africa: Sierra Leone.

Ecology: It is a demersal species.

Biology: The head of this species is well curved with rounded snout. Its mouth is small, and its lower lip is not developed. The eye is superolateral. The humeral process is elongate and pointed and in length it resembles pectoral fin. Barbels are with unilateral branching. Its beige-creamy body is mottled with reticulated dark brown markings. Dark bands are seen across the head. While its anterior part of the body is dark, posteriorly it is reticulate. A diagonal stripe is present from the rear of the adipose fin to the anal fin. The adipose fin has a light border and all other fins are barred and blotched. All barbels are also barred. It has a maximum length of 3.6 cm.

Venomous organs: The potential venom delivery system, which is the bony spine of this species, is Y shaped. No venom glands are visible in the spine (Wright, 2012).

Synodontis irsacae (Matthes, 1959)

Image credit: Larry M. Page (Reproduced with permission).

Common name(s): Upside-Down Catfish.

Global distribution: Africa: Lake Tanganyika.

Ecology: The fish inhabits muddy bottoms of the littoral zone and the benthic zones at a depth range of 0–40 m.

Biology: It has a strong, bony head capsule which extends up to the dorsal fin's first spine—a characteristic feature. Its head is with a humeral process which is a distinct narrow, bony, external protrusion. It has three pairs of barbels, namely, a pair of maxillary barbels which are located on the upper jaw, and two pairs of mandibular barbels of the lower jaw. Maxillary barbel extends up to the base of the pectoral fin. The outer pair of mandibular barbels extend only up to the front of the pectoral girdle. The body of the fish is with several skin folds which are tiny and vertical. Its mouth faces downward and it has wide lips containing papilla. The spine of the dorsal fin is short and slightly curved; and it is serrated only on its back. The spine of the pectoral fin is also slightly curved and is with large serrations on the back. The adipose fin is long and convex shaped. The caudal fin of this species is forked, with rounded lobes. The body coloration is grayish brown to copper brown above and underside is slightly lighter. The caudal fin is black with white at its end and barbels are white. This omnivorous fish feeds on a variety of diets such as sponges, crustaceans, gastropods, bivalves, insect larvas, eggs of other fishes, and algae. It has a maximum length of 19 cm.

Venomous organs: The potential venom delivery system, which is the bony spine of this species, is Y shaped. Paired anterior and posterior venom glands are seen (Wright, 2012).

Synodontis njassae Keilhack, 1908 (= *Synodontis zambesensis*)

Image credit: Paweł Cieśla Staszek_Szybki_Jest, Wikipedia. https://creativecommons.org/licenses/by-sa/4.0/

Common name (s): Njassae Synodontis Catfish; Malawi Squeaker; Upside-Down Catfish.

Global distribution: It is endemic to Lake Malawi, East Africa.

Ecology: It occurs in the inflowing streams and is known to be abundant on rocky shores.

Biology: It has characteristic small, medium, and large spots on its body. Its head is large and bony. It is a surface feeder preying only during night time on zooplankton. Its less-frequent food items include planktonic algae and insects and detritus. It attains a maximum length of 30 cm.

Venomous organs: The potential venom delivery system, which is the bony spine of this species, is Y shaped. Paired anterior venom glands are seen (Wright, 2012). The dorsal and pectoral fins possess a well-developed and serrated spine. The mucus covering these spines has been reported to be toxic (https://www.scotcat.com/articles/article119.html).

3.18 SHARK CATFISHES (PANGASIIDAE)

Helicophagus leptorhynchus (Ng and Kottelat, 2000)

Image credit: Panitvong, Nonn (Reproduced with permission).

Common name(s): Indonesian Catfish.

Global distribution: Tropical; Asia: Indochina.

Ecology: This demersal species occurs in permanent river channels. It is known to migrate upstream and downstream during the flood season and non-flood season, respectively.

Biology: The body of this species is long and laterally compressed. The snout is elongate and conical in dorsal view. Its mouth is narrow and its posterior nostril midway is located between anterior nostril and eye. In alcohol-preserved specimens, dorsum is gray. A small gray patch is present on its humeral region, immediately above the pectoral fin. The lower portions of the flanks and belly are whitish. The base of fins is dark yellow; the distal dorsal fin and caudal fin are black; and the distal regions of other fins are hyaline. It feeds largely on molluscan bivalves and it attains a maximum length of 47 cm.

Venomous organs: The potential venom delivery system, which is the bony spine of this species, is Y shaped. Spine has paired, hemispherical venom glands (Wright, 2012).

Pangasianodon hypophthalmus (Sauvage, 1878) (= *Pangasius sutchi*)

Image credit: Staticd Wikimedia Commons. https://creativecommons.org/licenses/by-sa/3.0/

Common name(s): Striped Catfish; Iridescent Shark; Shark Catfish.

Global distribution: Tropical; Asia: Chao Phraya and Mekong.

Ecology: This benthopelagic species inhabits large rivers it may also thrive in rice fields.

Biology: The body of this species is elongated and its head is relatively small. The mouth is broad and is in terminal in position. Its small sharp teeth are seen on jaw, vornerine, and palatal bones. Its eyes are large. Of its two pairs of barbels, the upper ones are shorter than the lower. Gill rakers are normally developed and small gill rakers are found interspersed with larger ones. Its fins are dark gray or black. The body of adult fishes is uniformly gray in color. This species has a dark stripe on the middle of its anal fin and a dark stripe in each of its caudal lobes. It has a maximum length of 130 cm.

Venomous organs and envenomation: The potential venom delivery system, which is the bony spine of this species, is Y shaped. Paired, hemispherical venom glands are seen in the spine (Wright, 2012). Human envenomation due to this species has been reported with the admission of a 51-year-old Malay man in the emergency department (Ahmad et al., 2013).

Pangasius bocourti (Sauvage, 1880)

Image credit: Ian Baird (Reproduced with permission).

Common name(s): Basa Catfish.

Global distribution: Tropical; Asia: Mekong and Chao Phraya basins.

Ecology: This benthopelagic species inhabits mainly large rivers However, it is also seen in deeper slower reaches. It is also known to enter flooded forest.

Biology: The body of this species is stout and heavy. Its head is broader than long. The dorsum is blueish black in color. The snout is blunt and is with broad white band on muzzle. A single large patch of vomerine teeth is seen on each side. It feeds mainly on plants and reaches a maximum length of 120 cm.

Venomous organs: The potential venom delivery system which is the bony spine of this species is Y shaped. Paired, hemispherical venom glands are seen in the spine (Wright, 2012).

Pseudolais pleurotaenia (Sauvage, 1878)

Image credit: Ian Baird (Reproduced with permission).

Common name(s): None.

Global distribution: Tropical; Asia: Mekong, Chao Phraya, Meklong, and Tapi basins.

Ecology: It is a benthopelagic; potamodromous species occurring in large and medium-sized rivers.

Biology: In this species, abdomen has a well-developed median keel which extends from throat to origin of anal fin. The dorsal fin is with a strongly serrated spine and its pectoral spines are slightly longer than dorsal spine. Pelvic fins are small and anal fin is low and long. The adipose fin is very small. A single, oval-shaped patch of teeth is seen on each side of palate, and the eyes are large. It feeds on terrestrial and aquatic insects and plant matter. Its maximum length is 35 cm.

Venomous organs: The potential venom delivery system, which is the bony spine of this species, is Y shaped. Paired, hemispherical venom glands are seen in the spine (Wright, 2012).

3.19 ANTENNAE CATFISHES (PIMELODIDAE)

Brachyplatystoma capapretum (Lundberg and Akama, 2005)

Image is not available.

Common name(s): Piraiba Catfish; Dark Caped Goliath Catfish.

Global distribution: Tropical; South America: Brazil and Peru.

Ecology: This benthopelagic species commonly inhabits muddy waters and deeper, flowing channels at a depth range, 5–10 m.

Biology: The adult fishes have dark grayish dorsum and plain white to dusky cream ventrum. Its jaw teeth are dimorphic with fine, outer teeth and slender, straight inner teeth. The body coloration is black to dark gray or brown dorsally and it is white ventrally. The caudal fin which is fairly forked with equal lobes bears a light patch. It attains a maximum length of 101 cm and weight of more than 30 kg. It is the South America's largest catfish. It is a carnivore with voracious feeding habits venomous organs: The shaft portion of the pectoral spines and the dorsal and anterior articular processes of the spine base are harmful. Further, its hunter type of teeth are also dangerous to humans. Owing to these features, this species is also known as man eater species (https://whatdewhat.com/dangerous-freshwater-fish-can-kill/).

Cheirocerus eques (Eigenmann, 1917)

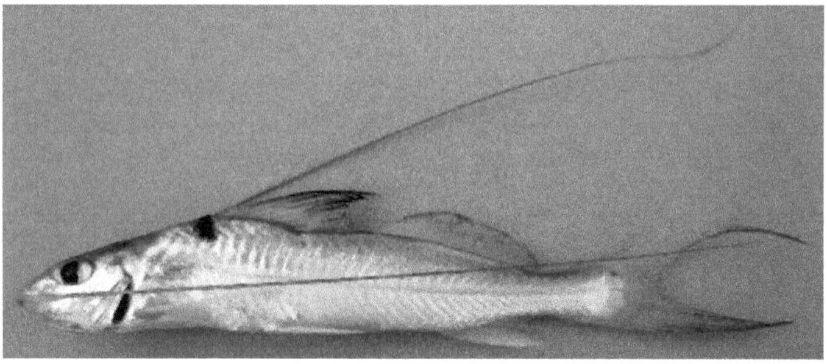

Image credit: Fernando M. Carvajal–Vallejos (Reproduced with permission).

Common name(s): Clarias Theodorae; Long-Whiskered Catfish.

Global distribution: Tropical; South America: Amazon River Basin.

Ecology: It occurs in slightly cooler and shallower rivers. Also found in sandy beaches in meter-deep, swiftly-flowing water.

Biology: In this species, its maxillary barbel extends beyond tips of the caudal fin; the outer mental barbel extends almost to tip of pelvic fin; and its inner mental barbel extends to the middle of depressed pectoral fin. The dorsal fin is relatively long and pointed. The adipose fin is long and fairly high. The caudal fin is deeply forked with long, pointed upper lobe; and its lower lobe is shorter and rounded at its tip. The anal fin extends posteriorly beyond posterior end of adipose-fin base when depressed. The pelvic fin reaches two-thirds distance to anal-fin origin, and depressed pectoral fin reaches posteriorly three-fourths of distance to pelvic-fin origin. The pectoral spine is slightly shorter than its first branched pectoral ray. The head and body are dusky gray above mid-lateral line and light colored ventrally. A distinct dark band crosses nape at dorsal-fin origin. The maxillary barbel is gray and mental barbels are unpigmented. The ventral lobe of caudal fin is gray and is dorsal lobe is lighter colored. The pectoral, pelvic, and anal fins are generally unpigmented. There are 17–21 gill rakers. It feeds mainly on benthic invertebrates and larvas of chironomids. It grows to a maximum length of 18 cm.

Venomous organs: The potential venom delivery system, which is the bony spine of this species, is Y shaped. Spine has no visible venom glands (Wright, 2012).

Hypophthalmus marginatus (Valenciennes, 1840)

Image credit: Jonathan Villamil Rodríguez (Reproduced with permission).

Common name(s): Long-Whiskered Catfish.

Global distribution: Tropical; South America: French Guiana and Suriname.

Ecology: It is a demersal, potamodromous species.

Biology: This species is characterized by its deep and large body. The head is flat and is much longer than wide. The snout is long. The dorsal fin is vertical and is slightly sloped at tip. Its ground coloration is light gray with blue gray on dorsum and head. The flanks and ventrum are white. The fins possess body coloration except caudal fin, which is blueish gray with black distal edge. The caudal and anal may also have a reddish coloration. The barbels are black. It has zooplanktivorous feeding habits. It has a maximum length of 55 cm.

Venomous organs: The potential venom delivery system, which is the bony spine of this species, is Y shaped. Spine has no visible venom glands (Wright, 2012).

Megalonema platanum (Günther, 1880)

Source: Public domain

Common name(s): Long-Whiskered Catfish.

Global distribution: Subtropical; South America: Paraná River basin.

Ecology: It is a demersal species.

Biology: In this species, back of its head is slightly darkened. The snout is rounded and the eyes are large. The fins are clear. The first ray of the pectoral fin is ossified. It is an endangered species and it has a maximum length of 40 cm and weight of more than 380 g.

Venomous organs: The potential venom delivery system, which is the bony spine of this species, is N shaped. There are no discernible glandular cells (Wright, 2012).

Parapimelodus valenciennes (Lütken, 1874)

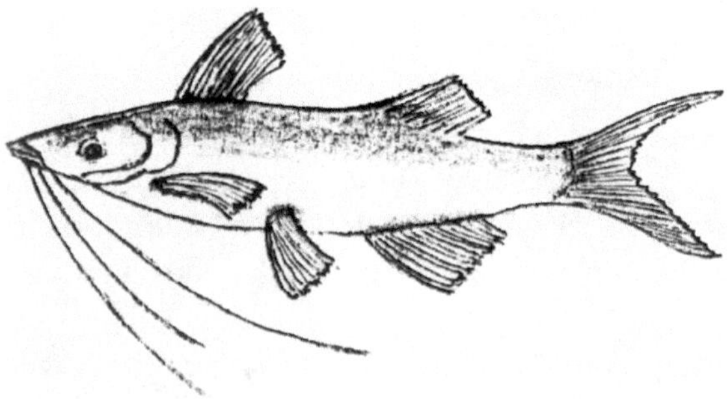

Imagecredit:https://enciclopediaanimal.wordpress.com/peces-de-la-cuenca-del-plata-%E2%80%93-parapimelodus-valenciennesi-%E2%80%A2-portenito/parapimelodus-valenciennesi-j/

Common name(s): None.

Global distribution: Tropical; South America: Paraná River basin.

Ecology: It is a demersal species.

Biology: It has extremely long gill rakers and more than 55 gill rakers are present on its first branchial arch, which is largely related to its plankton feeding habit. It has very large eyes on the sides of its head. It has a maximum length of 24 cm.

Venomous organs: The potential venom delivery system, which is the bony spine of this species, is Y shaped. Spine has no visible venom glands (Wright, 2012).

Pimelodus blochii (Valenciennes, 1840) (= *Pimelodus clarias*)

Image credit: Ictiologia Universidad Católica de Oriente. Wikimedia. https://creativecommons.org/licenses/by/2.0/deed.en

Common name(s): Bloch's Catfish.

Global distribution: Tropical; South America.

Ecology: This benthopelagic species thrives both in freshwater and brackish water biotopes. This gregarious fish may seek shelter in wooden crevices.

Biology: This species has a flat-fronted nose and long flowing barbels. It has its characteristic four lateral stripes on the flanks in its juvenile and these stripes may disappear in adults and it is becoming a duskier gray. Like the other members of the Pimelodidae family, it locates its food by smell. It mainly feeds on bloodworms and earthworms. It is also known to feed on fruits, small fishes, and insects and detritus. It is capable of intestinal breathing. It may migrate upstream during reproduction. It attains a maximum length of 35 cm.

Venomous organs: It is a venomous species and it has caused human envenomations from South America to Mexico (Dart, 2004).

Pimelodus maculatus (Lacepede, 1803)

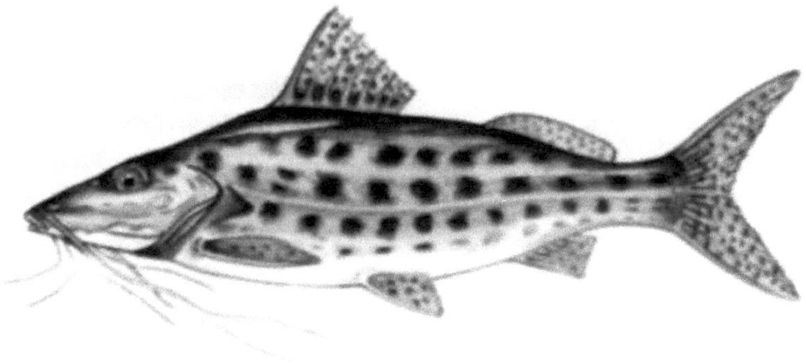

Image credit: Secretaria de Agricultura e Abastecimento. Wikimedia. https://creativecommons.org/licenses/by/2.0/deed.en

Common name(s): Spotted Pim.

Global distribution: Subtropical; South America.

Ecology: It is a benthopelagic, potamodromous species occurring mainly in ponds, pools, and small lakes. It is also seen on the sandy or muddy bottoms in shallow areas.

Biology: The entire surface of this fish is covered with large blotches—a characterized feature of this species. Males are smaller and less heavy than females. It is an omnivorous species consuming algae; larvas and adults of insects and crustaceans; mollusks and fish. It attains a maximum length of 51 cm and weight of 2.4 kg.

Venomous organs and envenomation: This species has been reported to be the most common cause of injuries in Brazil's lakes and rivers (de Aquino et al., 2016; Wikipedia). The potential venom delivery system, which is the bony spine of this species, is Y shaped. Paired anterior and posterior venom glands are seen in spine (Wright, 2012).

This species has been reported to cause accidents very frequently with the fishermen (Sarmiento et al., 2015). Its crude venom which is about 100 g of protein per fish is said to be involved in the human envenomation. Further, its venom is capable of inducing marked edematogenic and nociceptive effects; and heart problems at a dose of 30–100 g per experimental animal. Furthermore, this crude venom also induced marked alterations in the level of creatine kinase (CK) and its isoenzyme creatine kinase myocardial band (CK-MB) in mice.

Pimelodus pictus (Steindachner, 1876)

Image credit: Claire H., Wikipedia. https://creativecommons.org/licenses/by-sa/2.0/

Common name(s): Pictus Cat.

Global distribution: Tropical; South America: Amazon and Orinoco River basins.

Ecology: This benthopelagic species is found in shallow, flowing waters over sandy or muddy substrates.

Biology: This species has extremely long barbels which can extend up to its caudal fin. It is silver-colored with black spots and stripes. Sharp spines are seen on its dorsal and pectoral fins. It has a downturned mouth and a forked tail. In this species, there is little sexual dimorphism, with larger females than males. It is an active swimmer and nocturnal bottom-feeder. Two color forms appear in this species. The first is Peruvian variety that has large "dalmatian" style spots all over its body and smaller spots around the head. The other form is Colombian variety that grows larger than the first form has many smaller spots which are of pin–prick size. It has a maximum length of 11 cm.

Venomous organs: The spine of its dorsal fin is mildly venomous (https://en.wikipedia.org/wiki/Pimelodus_pictus). Hardy et al. (2016) stated that this species may do harm due to its pectoral-fin spine. The leading-edge fin ray of its pectoral fin is modified into a hardened spine with serrated dentations along both edges.

Pseudoplatystoma corruscans (Spix and Agassiz, 1829)

Source: Public domain

Common name(s): Spotted Sorubim; Spotted Catfish.

Global distribution: Subtropical; South America: São Francisco and Paraná River basins.

Ecology: This demersal, potamodromous species Inhabits major water-courses in deep channels of major rivers.

Biology: The body of this species is somewhat stout. Lateral sides of its head are almost parallel. The top and sides of the body and the head are with dark brownish pigmentation in the form of spots distributed in rows. A row of spots may also be seen below lateral line. Pale vertical bars are found on lateral region of body from opercular flap to caudal peduncle. The dorsal fin rays possess small spots. The pectoral fin is dark and without spots. Pelvic fin is pale and anal fin is pale, dark, or spotted. It feeds mainly on other fishes and attains a maximum length of 166 cm and weight of about 100 kg.

Venomous organs: Large, serrated barbs present on its fins have been reported to cause lacerated wounds, edema, erythema, and skin necrosis (de Aquino et al., 2016).

Pseudoplatystoma fasciatum (Linnaeus, 1766)

Image credit: Matthieu Sontag. Wikimedia Commons. https://creativecommons.org/licenses/by-sa/3.0/

Common name(s): Barred Shovelnose; Barred Sorubim; Barred Catfish; Tiger Catfish.

Global distribution: Colombia, Ecuador, Venezuela, Guyana, Suriname, Bolivia, Brazil, French Guiana, Peru, and Argentina.

Ecology: It inhabits a variety of biotopes, namely, large rivers, lakes, and flooded forest. It is often found in more sluggish habitats.

Biology: In this species, its posterior dorsal fin is adipose. The anal fin is with 13 rays; and pelvic fins are with six rays and spotted. The dorsal fin is with seven rays; and the caudal fin is with 17 rays. The lower jaw is short; and rostrum is flat and dark. Its head is long, flat, and rounded anteriorly. The dorsum is with black and white stripes on each side. Its abdomen is white. All fins are sprinkled with black spots. While the dorsal fin is near the head, adipose is near the caudal fin. It reaches a maximum length and weight of 104 cm and 70 kg, respectively.

Food and feeding: It is a nocturnal predator feeding mainly on other fishes, crabs, prawns, mussels, cockle, or earthworms.

Reproduction: It is known to have long distance migrations for spawning. Females and males are known to sexually mature at 56 and 45 cm, respectively. Its fecundity is about 8 million eggs per kg; hatching takes place in 24 h after fertilization. Larvas begin carnivorous feeding 2 days after hatching.

Venomology: The crude venom of this species has shown strong nociceptive and edema effects in mice (Sarmiento et al., 2015). Aquino et al. (2016) reported that this species has caused injuries in the professional fishermen in the Corumbá and Miranda towns in the Pantanal region.

Pseudoplatystoma reticulatum (Eigenmann and Eigenmann, 1889)

Image courtesy: Chucao, Wikipedia. https://creativecommons.org/licenses/by-sa/3.0/

Common name(s): Striped Catfish; Reticulated Shovelnose.

Global distribution: Subtropical; South America: Argentina, Bolivia, Brazil, Paraguay and Uruguay.

Ecology: This demersal species inhabits major watercourses in deep channels of major river systems.

Biology: The body of this species is normally stout. The head of this species is somewhat pointed on the anterior margin when it is viewed dorsally. The dorsal lip expands toward the midline. A dark pigmentation is seen on dorsal and lateral region of body. Dark loop–like bars are seen all over the dark area of the body from posterior opercular flap to caudal peduncle. Its head is also with spots and bars from its tip to nape. The dorsal-fin spine and rays are provided with spots. The anal fin is spotted. The pectoral and pelvic fins are dark dorsally, and pale ventrally and are with few spots. The dusky posterior margins may also be seen on pectoral fins of some individuals. The adipose fin is with few spots. While the area nearer to the dorsal procurrent caudal rays is with spots, the ventral procurrent caudal rays are without spots. It attains a maximum length of 60 cm.

Venomous organs: Large, serrated barbs present on its fins have been reported to cause lacerated wounds, edema, erythema, and skin necrosis (de Aquinol et al., 2016).

Sorubim lima (Bloch and Schneider, 1801)

Image credit: Sitron, Wikimedia. https://creativecommons.org/licenses/by-sa/3.0/

Common name(s): Duckbill Catfish.

Global distribution: Tropical; South America: Brazil, Ecuador, Peru, and Bolivia.

Ecology: This demersal species occurs in schools and is mainly nocturnal.

Biology: The body of this species is elongate, anteriorly cylindrical and is compressed posteriorly. Its head is strongly depressed. The snout is spatulate and is overhanging the mouth. Its fins are short. The caudal fin is deeply forked and its upper lobe is produced to a point and lower lobe is rounded. It has three pairs of barbels. The body coloration is uniform silver gray and the belly is of pure white. Fins are colorless and caudal fin is with a dark center. It feeds mainly on fishes and crustaceans. It attains a maximum length of 54 cm and weight of 1.3 kg.

Venomous organs: The potential venom delivery system, which is the bony spine of this species, is Y shaped. Paired anterior and posterior venom glands are seen in spine (Wright, 2012).

Zungaro zungaro (Humboldt, 1821) (= *Pseudopimelodus zungaro*)

Source: Public domain

Common name(s): Gilded Catfish; Black Manguruyu.

Global distribution: Tropical; South America: Amazon and Orinoco River basins.

Ecology: This benthic species is mainly seen in calm upstreams of big tributaries which are with muddy bottoms.

Biology: It is an active predator and scavenger preying on smaller fish, invertebrates, and carrion. It attains a maximum length of 140 cm and weight of 50 kg.

Venomous organs: The potential venom delivery system, which is the bony spine of this species, is Y shaped. Paired, small, and anterior venom glands are present (Wright, 2012).

3.20 EELTAIL CATFISHES (PLOTOSIDAE)

Neosilurus ater (Perugia, 1894)

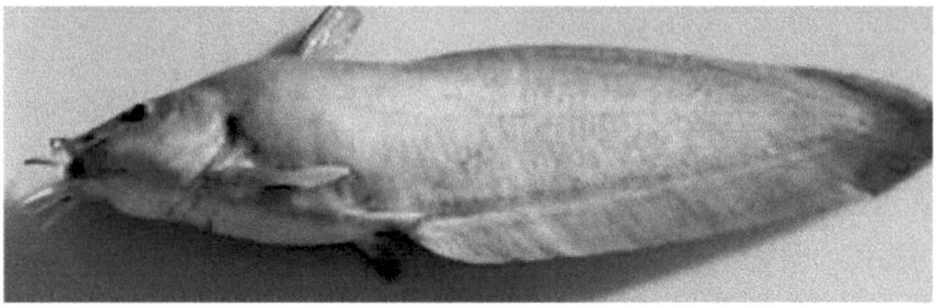

Image credit: Alan Webb, James Cook University (Reproduced with permission).

Common name(s): Narrowfront Tandan.

Global distribution: Tropical; Asian and Oceanic countries such as Australia and New Guinea.

Ecology: This solitary, demersal species is found in rivers and smaller streams with strong currents. It also lives in quiet or week flowing waters such as side channels and pools.

Biology: It is normally identified by its elongate snout. Its body may have a mottled pattern and its coloration is gray to blackish and belly is pale. In some individuals, both males and females develop a bright golden-orange color on their sides especially during breeding season. Its mouth is surrounded by four pairs of barbels. It consumes mainly worms, crustaceans, mollusks, and insects. It has a maximum length of 47 cm and weight of 2 kg.

Venomous organs: It is a venomous catfish (https://www.fishbase.se/summary/Neosilurus-ater). Its dorsal and pectoral fins which are armed with serrated venomous spines may inflict a very painful sting (https://fishesofaustralia.net.au/home/species/2762).

Neosilurus brevidorsalis (Günther, 1867)

Image credit: Allan James (Reproduced with permission).

Common name(s): Short Finned Tandan; Short-Finned Catfish.

Global distribution: Asia and Oceania: Northern Australia; Central–Southern New Guinea.

Ecology: This demersal species inhabits lakes, swamps, rivers and backwaters. It also lives in flooded lagoons and billabongs.

Biology: The body coloration is dark brown to gray dorsally and belly is white. It consumes mainly prawns, mollusks, small crayfish, and insects. It attains a maximum length of 20 cm.

Venomous organs: It is a venomous species doing harm to intruders (https://www.fishbase.in/summary/12095).

Plotosus papuensis (Weber, 1910)

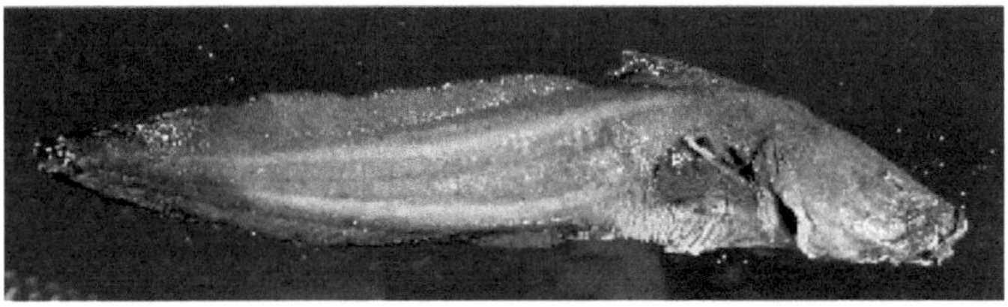

Image credit: Sabaj Perez, Mark (Reproduced with permission).

Common name(s): Papuan Eel Catfish.

Global distribution: Tropical; Asia and Oceania: New Guinea.

Ecology: This species is found in turbid rivers, backwaters, and swampy lagoons.

Biology: This species has an elongated body which is compressed and tapering toward the end of caudal fin. While its first dorsal fin bears strong serrated spines, its second dorsal fin is long and is confluent with the tail fin. The pectoral fin is also with strong and serrated spine. Anal fin is also more or less confluent with its caudal fin. A complete lateral line is present. It attains a maximum length of 100 cm.

Venomous organs: The first dorsal and pectoral fins that are associated with production of venom may harm the people (Usman et al., 2013; Anon., https://www.fishbase.de/summary/plotosus-papuensis.html).

Porochilus argenteus (Zietz, 1896)

https://fishesofaustralia.net.au/home/species/2767
Image credit: Dave Wilson / Aquagreen Museums Victoria https://creativecommons.org/licenses/by/3.0/au/

Common name(s): Silver Tandan.

Global distribution: Tropical; Oceania: endemic to Australia.

Ecology: This benthic species which is largely associated with sand or rock bottoms is common in waterholes in larger rivers and slow-flowing streams.

Biology: It is a slender bodied fish with a concave head-snout profile. Its eyes are relatively low and are closer to the mouth. Its head is much steeper. The caudal fin is fairly pointed. Serrations are seen on the inner side of the pectoral spine. The body coloration is pale silvery-white and fins are yellowish orange. This species is sexed by its urogenital papilla. While it is short and flat in females, it is longer and pointed in males. It has a maximum length of 35 cm.

Venomous organs: It is a venomous species (https://www.fishbase.in/summary/Porochilus-argenteus.html).

Porochilus rendahli (Whitley, 1928)

Image credit: Dave Wilson. Museums Victoria https://creativecommons.org/licenses/by/3.0/au/

Common name(s): Rendahl's Catfish.

Global distribution: Oceania: endemic to Australia.

Ecology: This freshwater, demersal species occurs in billabongs and streams.

Biology

Description: It is distinguished by its concave head which rises ahead of the dorsal fin thus giving this fish a hump back appearance. Its eyes are located closer to the tip of the snout. Body coloration may vary from a mottled light gray, to smooth dark color. A light gold/tan hue may also be seen in some individuals. It attains a maximum length of 24 cm.

Food and feeding: It feeds aquatic invertebrates and microcrustaceans including chironomid midge larvas and small ephemeropteran nymphs.

Reproduction: Reproduction of this species is not fully understood. Male matures at 10 cm and female at 11 cm. Rising temperatures and onset of wet season have been reported to trigger migration of this species to spawning areas, namely, muddy lowland lagoons. Eggs laid by the females measure 1.3 mm in diameter.

Venomology: It is a venomous species (https://fishesofaustralia.net.au/home/species/2768).

Tandanus bostocki (Whitley, 1944)

Image credit: Gerald R. Allen / Western Australian Museum. Museums Victoria https://creativecommons.org/licenses/by/3.0/au/

Common name(s): Freshwater Cobbler.

Global distribution: Oceania: endemic to Australia.

Ecology: This freshwater, demersal species inhabits slow-flowing streams, reservoirs, and ponds. It may also occur in lakes or pools in riverbeds. This species has also been reported to take shelter in underwater cavities in or near river banks and root mounds. It is tolerant of brackish water conditions.

Biology: This species is distinguished by its large, eel-like tail, that is, second dorsal fin is united with the anal and tail fins. It has a golden brown or dark brown coloration. Its belly, however is pale. It also shows a mottled or spotted pattern. It attains a maximum length of 90 cm.

Food and feeding: It consumes, shrimp, crayfish, molluscs, and juveniles of fish.

Reproduction: In this species, females mature at about 17 cm and November–December is the peak spawning season of this fish especially in the Blackwood River.

Venomology: This species has a venomous spine (https://www.fishbase.in/summary/12099; https://en.wikipedia.org/wiki/Freshwater_cobbler). It has caused human envenomations in Indo-pacific region and Australia https://rivers.dwer.wa.gov.au/species/tandanus-bostocki/#:~:text=The%20mouth%20is%20surrounded%20by,venom%20gland%20cells%20is%20torn reported that both pectoral fins and first dorsal fin of this fish have only one venomous spine and its venom may cause extreme pain. It is also believed that the venom of this species may cause localized edema due to a hemolytic effect. Other symptoms include increased blood flow in the injured area and vasoconstriction.

Tandanus tandanus (Mitchell, 1838)

Image credit: Fir0002/Flagstaffotos; Wikimedia. https://www.gnu.org/licenses/old-licenses/fdl-1.2.html

Common name(s): Freshwater Catfish; Eel-Tailed Catfish; Tandan Catfish.

Global distribution: Temperate; Oceania: endemic to Australia.

Ecology: This demersal and potamodromous species inhabits ponds, lakes, and slow-moving streams with vegetation. This species, which does not form schools, swims close to gravel or sand bottoms and is believed to be abundant in lakes.

Biology: In this species, the head is large and is with fleshy and thick lips and tubular nostrils. There are four pairs of barbels in its down-turned mouth. It lacks scales. The first dorsal fin is fairly high, with only serrated spine. Its skin is tough and smooth. The dorsum may be of varying colors such as black brown, purplish, or olive green and it is white ventrally. Urogenital papilla is triangular in females; and it is longer and cylindrical in males. It is mainly a bottom-feeder on crayfish, prawns, mollusks, and small fishes. It may also consume insect larvas. It attains a maximum length of 90 cm and weight of 7 kg.

Venomous organs: The first dorsal and pectoral fins are preceded by a sharp serrated spine which may be harmful to intruders (https://www.fishbase.se/summary/Tandanus-tandanus).

Tandanus tropicanus (Welsh et al., 2014)

Image credit: Professor Dean Jerry and Professor Damien Burrows (Reproduced with permission).

Common name(s): Eeltail Catfish.

Global distribution: Oceania: Australia.

Ecology: This freshwater, demersal species inhabits coastal rivers.

Biology: It has an elongate and robust adult body. The body profile is slightly convex between posterior serrae and origin of first dorsal fin. The posterior half of the body is progressively compressed toward caudal end with a continuous caudodorsal and anal fin. Its head is broad and is slightly depressed. The eyes are moderate in size and are positioned dorsolaterally. Mouth is subterminal to slightly inferior in position. Lower and upper lips are thick and fleshy. It attains a maximum length of 40 cm.

Food and feeding: It is a carnivore feeding largely on benthic invertebrates.

Reproduction: Not reported.

Venomology: Its venomous pectoral spine has serrations on its posterior edge often in two longitudinal rows (Welsh et al., 2014; https://fishesofaustralia.net.au/home/species/5006).

3.21 GLASS CATFISHES (SCHILBIDAE)

Laides longibarbis (Fowler, 1934)

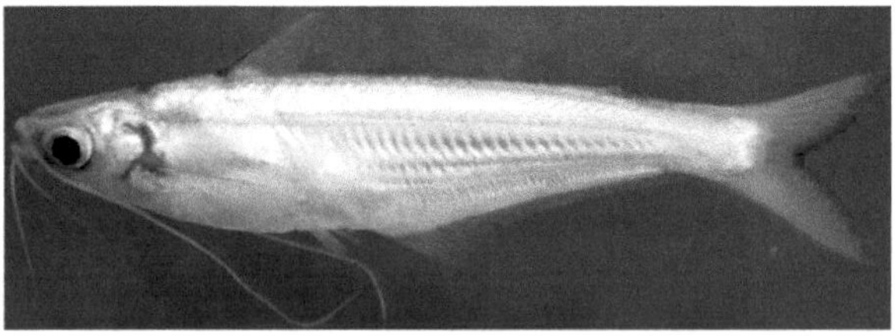

Image credit: Chavalit Vidthayanon (Reproduced with permission).

Common name(s): Not reported.

Global distribution: Tropical; Asia: Mekong, Chao Phraya and Meklong basins.

Ecology: This demersal species inhabits large rivers with turbid and or standing waters.

Biology: It has a smaller eye and a longer anal fin base. Ribbon-shaped mandibular and mental pairs are seen on its lower jaw. It feeds on zooplankton and fishes. It has a maximum length of 14 cm.

Venomous organs: The potential venom delivery system, which is the bony spine of this species, is Y shaped. Spine has small, paired, posterior visible venom glands (Wrihgt, 2012).

Pseudeutropius brachypopterus (Bleeker, 1858)

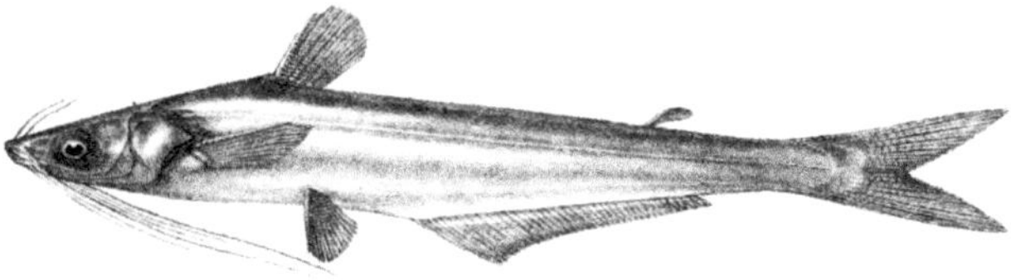

Pseudeutropius sp.

Source: Public domain

Common name(s): Mini Shark Cat; Dwarf Pangasius.

Global distribution: Tropical; Asia: Sumatra and Borneo.

Ecology: This demersal species is mainly found in rivers and lakes.

Biology: Its eight long barbels are located around the mouth. According to the need, these barbels are either attached to the body or stretched in front of the head. Further, it stretches its barbel in a jerky manner as putting up an umbrella. It has a much shorter anal fin and it feeds mainly on insect larvas and small invertebrates and attains a maximum length of 12 cm.

Venomous organs: The potential venom delivery system, which is the bony spine of this species, is Y shaped. Spine has paired posterior visible venom glands (Wright, 2012).

Schilbe mystus (Linnaeus, 1758)

Image credit: Dr. Frank Teigler (Reproduced with permission).

Common name(s): African Butter Catfish.

Global distribution: Tropical; West Africa: Senegal, Cross, Wouri, and Sanaga Rivers.

Ecology: This demersal species is lacustrine as well as fluviatile with potamodromous migration.

Biology: In this species, the mouth is subterminal and the snout is reaching beyond lower jaw. The anterior nostrils are closer to each other. The nasal barbel reaches anterior eye margin. Its lower limb of first gill arch bears 8–15 gill rakers. The adipose fin is fully developed. The inner margin of pectoral-fin spine is strongly denticulated posteriorly. It has a brownish coloration on its head and dorsal surface and silvery white on the belly. The fins are usually colorless. It has been reported to feed from surface waters and mid-water on a wide variety of food items including insects, shrimps, snails, fish, etc. It has a maximum length of 35 cm and weight of 250 g.

Venomous organs: The potential venom delivery system, which is the bony spine of this species, is Y shaped. Spine has large hemispherical visible venom gland (Wright, 2012).

Siluranodon auratus (Geoffroy Saint-Hilaire, 1809)

Image credit: Dr. Abu-Almaaty (Reproduced with permission).

Common name(s): Butterfish.

Global distribution: Tropical; Africa: Chad, Niger, Volta, Comoe, and Nile.

Ecology: This pelagic fish is demersal and potamodromous. It occurs mainly in large and moderate lakes and rivers. It may also be seen beneath fringing *Eicchornia* vegetation.

Biology: It has a silvery head and brown back. A large brown stripe is seen along the lateral line. Similarly, a large brown blotch is present behind its head and from this blotch a second stripe also starts to continue above the base of the anal fin. A dark line above the base of the anal fin joins this second stripe at caudal peduncle while another brown stripe borders the anal fin over its complete length. Other fins are weakly speckled in brown. However, some populations may be all black with little silvery tinge between the stripes. A red spot with a golden tinge is also present on the gill cover at maturity. There are about 60 thin and flexible gill rakers on lower limb of first gill arch; and four pairs of fully developed barbels. The dorsal fin is small and spineless. Its jaws and palatine bones are without teeth. It is a microphagous feeder consuming mainly algae and occasionally on rotifers, planktonic and benthic crustaceans; insects; and debris. It may attain a maximum length of 14 cm.

Venomous organs: The potential venom delivery system, which is the bony spine of this species, is Y shaped. Spine has no visible venom glands (Wright, 2012).

3.22 SHEAT CATFISHES (SILURIDAE)

Belodontichthys truncatus (Kottelat and Ng, 1999)

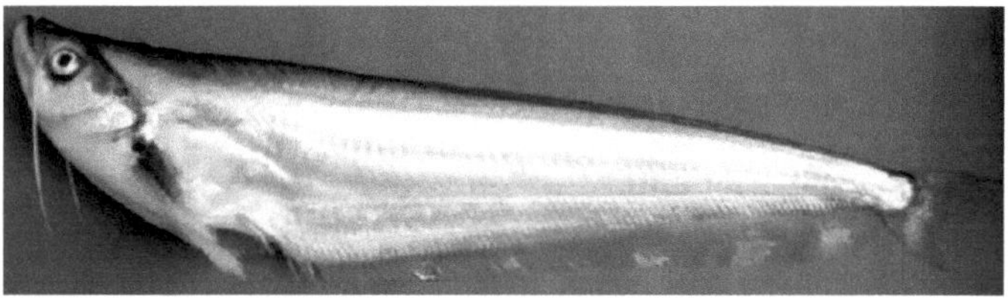

Image credit: Chavalit Vidthayanon (Reproduced with permission).

Common name(s): Dracula Sheatfish.

Global distribution: Tropical; Asia: Cambodia, Vietnam, Thailand and Laos.

Ecology: This demersal species occurs at surface and middle depths in medium-to-large-sized rivers.

Biology: It has a strongly compressed head and body; and an upturned mouth with large teeth. Its pectoral fin is longer than head. It feeds mainly on smaller fish near the surface. It attains a maximum length of 60 cm.

Venomous organs: The potential venom delivery system, which is the bony spine of this species, is N shaped. There are no discernible glandular cells (Wright, 2012).

Kryptopterus bicirrhis (Valenciennes, 1840)

Image courtesy: Caroline, ENCYCLO FISH (Reproduced with permission).

Common name(s): Glass Catfish.

Global distribution: Tropical; Asia: Malay Peninsula, Sumatra, and Borneo.

Ecology: It inhabits large rivers with turbid water, where they normally live near the shore. It is known to enter flooded fields and is also found in lowland streams. It is diurnal and it may school up to 100 fish.

Biology: In this species, most of its body is not clearly transparent and it is only transparent around the head. The body is mostly opaque and light grayish. In light, its body has an iridescence, and it becomes milky white when it dies. Its two long barbels are as long as the body. The dorsal fin is very much reduced to a small triangle structure, and its pectoral fins are longer than the head. It is a diurnal predator feeding mostly on worms, crustaceans, insects and small fishes. It attains a maximum length of 15 cm.

Food and feeding: It is omnivore feeding mainly on artemia and mosquito larvas.

Reproduction: There is not much information on its breeding.

Venomous organs: The potential venom delivery system, which is the bony spine of this species, is Y shaped. Spine has posterior, visible glandular cells (Wright, 2012).

Pterocryptis berdmorei (Blyth, 1860)

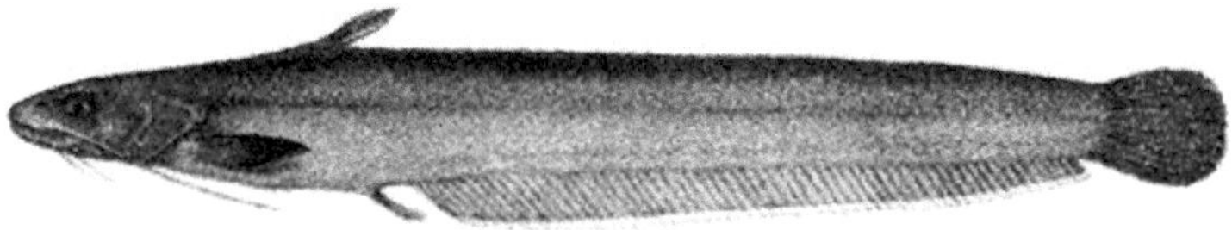

Pterocryptis sp.

Source: Public domain

Common name(s): Burmese Silurus.

Global distribution: Tropical; Asia: Myanmar, India, and Western Thailand.

Ecology: It is a demersal species.

Biology: In this species, there are 15–16 dorsolateral white dot sensory pores which are arranged in equal intervals from tip of snout to base of caudal fin. There are also five sensory pores which form an oblique line above operculum on each side of head. It has a small dorsal fin and its upper jaw is longer than the lower jaw. Its anal fin is confluent with tail fin and there is a distinct notch between them. It has a maximum length of 21 cm.

Venomous organs: The potential venom delivery system, which is the bony spine of this species, is Y shaped. Paired and large posterior venom glands are seen in the spine (Wright, 2012).

Pterocryptis cochinchinensis (Valenciennes, 1840)

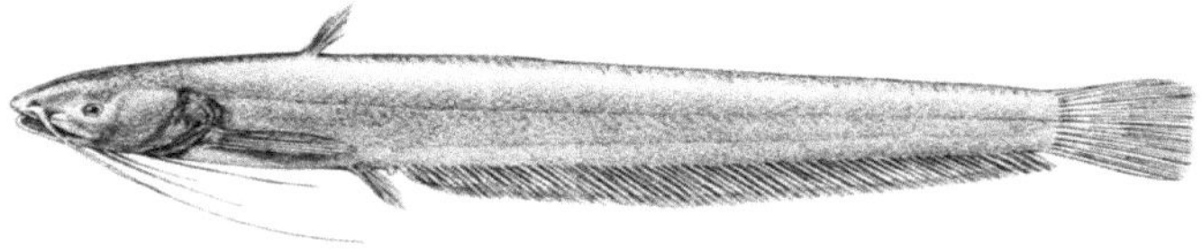

Source: Public domain

Common name(s): None.

Global distribution: Tropical; Asia: Laos, Vietnam, Thailand, and Southeastern China.

Ecology: This demersal species inhabits mainly streams.

Biology: It has a dark brown body; dorsal surface is purple; stomach is light gray; and all fins are light gray. Its eye is circular. The presence of a large genital papilla is the characteristic feature of this species. While males are with a long conical papilla, females are with a flat, leaf-shaped papilla with a small projection at the tip. It feeds actively at night on small fish, crustaceans, and insects. It has a maximum length of 40 cm and weight of 430 g.

Venomous organs: The potential venom delivery system, which is the bony spine of this species, is Y shaped. Paired and large posterior venom glands are seen in the spine (Wright, 2012).

Pterocryptis taytayensis (Herre, 1924) (= *Hito taytayensis*)

Image credit: Dr. Joie Matillano (Reproduced with permission).

Common name(s): Not designated.

Global distribution: Tropical; Asia: Palawan, Philippines.

Ecology: It is a demersal species.

Biology: In this species, its upper jaw is longer than the lower jaw. The dorsal and pelvic fins are very small. The pectora fins are short and their spine is stout and sharp-pointed. The anal fin is free from the caudal fin and is long and wide. Preserved specimens are yellowish brown. A large dark lateral blotch is seen between pectoral fin and lateral line. Fins are colorless. It has a maximum length of 12 cm.

Venomous organs: The potential venom delivery system, which is the bony spine of this species, is Y shaped. Large and posterior venom gland is seen (Wright, 2012).

Silurichthys schneideri (Volz, 1904)

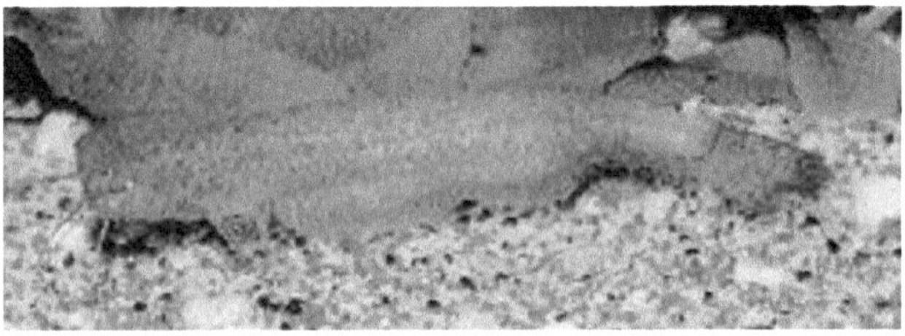

Image credit: วัตุลึกลับใต้น้ำ Wikipedia. https://commons.wikimedia.org/wiki/GNU_Free_Documentation_License

Common name(s): None.

Global distribution: Tropical; Asia: Indonesia, Malaysia, Thailand, and Cambodia.

Ecology: This benthic species is commonly seen in fast-flowing mountain streams with rocky or sandy bottoms. It may also be seen hiding under rocks.

Biology: The body of this species is long and is laterally compressed. There are two to four gill rakers on its first branchial arch. The caudal fin is emarginate and its upper lobe is 1–2 times longer than lower lobe. The pectoral spine and articulated segments are sexually dimorphic especially in mature individuals. In alcohol-preserved specimens, flanks are pale brown with weak randomly distributed irregular dark patches. The dorsal surface and sides of head are brown. Ventral surfaces of head, breast, and belly are yellow colored with scattered melanophores. The maxillary and the mandibular barbels are brown. The anal fin has irregular dark-brown patches and a hyaline ventral margin. The caudal fin is with irregular dark-brown patches and a hyaline posterior margin. The pectoral and pelvic fins are hyaline. The pectoral fin may have small dark-brown spots occasionally. It may attain a maximum length of 16 cm.

Venomous organs: The potential venom delivery system, which is the bony spine of this species, is Y shaped. Paired and large posterior venom glands are seen in the spine (Wright, 2012).

Silurus asotus (Linnaeus, 1758)

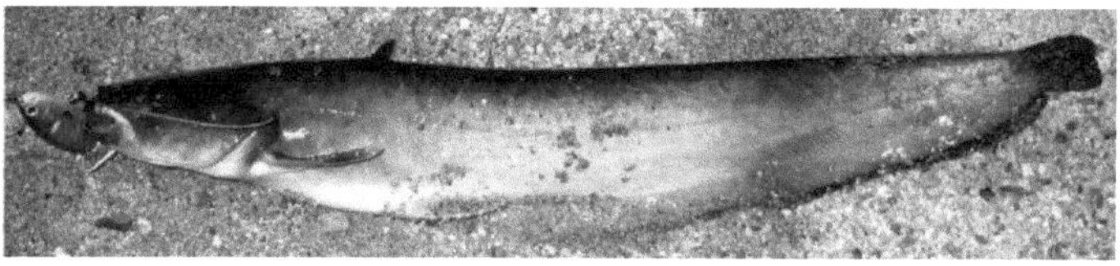

Image credit: Rhosoi, Wikipedia. https://creativecommons.org/licenses/by/2.0/

Common name(s): Amur Catfish.

Global distribution: Temperate; Asia: Korean Peninsula, Taiwan, Japan, China, and Russia.

Ecology: This demersal. Species exists in slow-flowing, medium sized rivers, lakes, and irrigation canals with muddy bottom.

Biology: The body of this species is elongated and cylindrical. The coloration of body is dark gray dorsally; belly is white; and sides are with irregular white spots. One pair of the maxillary barbel is longer than the head, and one pair of mandibular barbels is only about one-fifth of the length of the maxillary barbel. The dorsal fin is small and anal fin is connected with the caudal fin. The adult fishes feed on a variety of other fishes and attain a maximum length of 130 cm and weight of 30 kg.

Venomous organs: The potential venom delivery system, which is the bony spine of this species, is Y shaped. Spine has paired anterior visible venom glands and posterior visible glandular cells (Wright, 2012).

Silurus mento (Regan, 1904)

Image is not available.

Common name(s): Kunming Catfish.

Global distribution: Asia: endemic to Dian Chi Lake, Yunnan, China.

Ecology: This demersal species lives in the edge of lakes with plenty of aquatic vegetation where it is known to stay at bottom of water during the day and it comes out to the surface for seeking food during dawn and dusk.

It depends largely on aquatic macrophytes for shelter and its absence may result in increased mortality from its natural enemies including human beings.

Biology: The head of this species is depressed; its eyes are small and are located in anterior part of head; the snout is short; the lower jaw is longer than the upper one and its end is extending to almost the anterior edge of eyes; and barbels are in two pairs. The body is naked and is dark gray on back and sides, with cloudy streaks. Its belly is milky white. It is a carnivorous species. It is attaining a maximum weight of 38 cm.

Venomous organs: The potential venom delivery system, which is the bony spine of this species, is Y shaped. Spine has no visible venom glands (Wright, 2012).

Wallago micropogon (Ng, 2004)

Image credit: Heinz (Reproduced with permission.)

Common name(s): Helicopter Catfish.

Global distribution: Tropical; Asia: Vietnam, Laos, and Thailand.

Ecology: This demersal species is known to enter the flooded forest areas especially during high water conditions and remain at the edge of the forest during low water. It may also migrate into smaller streams for the purpose of spawning.

Biology: The body of this species has dark coloration with a mottled pattern of pale spots. The first branchial arch bears 14–16 gill rakers. A small gape extends to anterior margin of the eye; the eyes are small. The maxillary barbels reach to middle of pectoral-fin base. The pectoral fin is rounded. Its pectoral spine length is about 15% of its standard length. The adult fishes feed strictly on fishes. It attains a maximum length of 154 cm.

Venomous organs: The potential venom delivery system, which is the bony spine of this species, is Y shaped. There are no discernible glandular cells (Wright, 2012).

KEYWORDS

- **common name**
- **global distribution**
- **ecology**
- **biology**
- **venomous organs**

CHAPTER 4

BIOLOGY AND ECOLOGY OF VENOMOUS MARINE CATFISHES

ABSTRACT

The aspects of biology and ecology of venomous seawater catfishes belonging to a total of seven families of the Order, Siluriformes viz. Ailiidae, Ariidae, Bagridae, Doradidae, Plotosidae, Schilbeidae, and Siluridae are dealt with in this chapter.

4.1 ASIAN BUTTER CATFISHES (AILIIDAE)

Ailia coila (Hamilton, 1822)

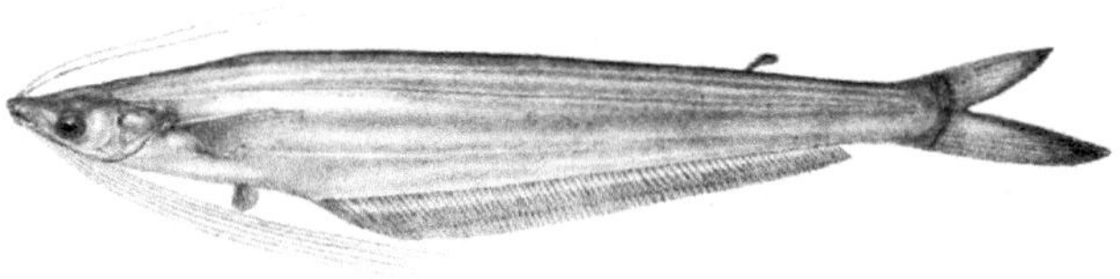

Source: Public domain

Common name (s): Gangetic ailia.

Global distribution: Tropical Asia: India, Pakistan, Bangladesh, and Nepal.

Ecology: This pelagic species inhabits surface to mid-waters of large rivers. It may form shoals and enter brackish waters also.

Biology: The abdomen of this species is rounded. There are four pairs of barbels surrounding the mouth. A short, insignificant adipose fin is present and rayed dorsal fin is absent. A slender pectoral fin with serrated spine

Biology and Ecology of the Venomous Catfishes. Ramasamy Santhanam, PhD (Author)

is present. Pelvic fin is small with six rays. Anal fin is considerably long and tail fin, which has a much longer lower lobe, is well marked. There is a simple and complete lateral line. Body is silvery and its fins are grey. Bases of anal and caudal fins are slightly yellowish. Occiput is with black spots. It reaches a maximum length of 30 cm.

Venomous organs: The potential venom delivery system, which is the bony spine of this species, is Y-shaped. There are no visible venom glands in the spine (Wright, 2012).

Clupisoma garua (Hamilton, 1822)

Image credit: Dr. Mostafa Hossain (Reproduced with permission).

Common name (s): Garua bachcha.

Global distribution: Tropical; Asia: Pakistan, India, Bangladesh, and Nepal.

Ecology: This demersal and potamodromous species inhabits large fresh water and tidal rivers. It may also be found in brackish waters.

Biology: Abdominal profile between pelvic and vent of this species is keeled. Head is oval and blunt; and snout is rounded. Occipital process does not reach the basal bone. Eyes are situated partly on the lower surface of the head and is with broad, circular adipose lid. Mouth is sub-terminal; teeth are in villiform bands on jaws; and vomeropalatine teeth are in a semilunar band, and are broken into four patches. Of its four pairs of barbels, nasal barbels do not reach the eye and maxillary barbels extend to the base of pelvic fins. Both the mandibular pairs extend to pectoral fins. Dorsal fin spine is slender, rugose anteriorly, and weekly serrated posteriorly; and adipose dorsal fin is absent. Pectoral fin is with a strong serrated spine. Anal fin is long and caudal fin is deeply forked. Body coloration is silvery grey on the back and lighter on its sides and abdomen. Fins are tinted grey. It is an omnivorous fish with bottom and marginal feeding habit. It attains a maximum length of 61 cm.

Venomous organs: The potential venom delivery system, which is the bony spine of this species, is Y shaped. Spine has no visible venom glands (Wright, 2012).

4.2 SEA CATFISHES (ARIIDAE)

Ariopsis felis (Linnaeus, 1766) (= *Galeichthys felis; Arius felis*)

Source: Public domain

Common name (s): Saltwater catfish, hardhead sea catfish, sea cat, and tourist trout.

Global distribution: Subtropical; Western Atlantic: Mexico to Florida and Massachusetts in USA; and Mexico.

Ecology: It is a marine, brackish, and reef-associated species. It is frequently seen in muddy bottoms and often enters estuaries.

Biology: Its hard bony plate located from between the eyes to the dorsal fin earns its common name. Its characteristic features include an elongated, naked body, a forked tail fin, a fairly flattened head, inferior mouth, and the presence of a broad arc in its upper jaw. Among its six barbels, two longer ones are present at the rear corners of the mouth and the remaining four barbels are seen under the chin. Dorsum of the fish is silvery with green, blue, or brownish tones. Belly is white. Dorsal and caudal fins are with dusky coloration. Pectoral, pelvic, and anal fins are either pale or yellowish. Females possess fairly larger pelvic fins than their counterparts. Its diet mainly consists of seaweeds seagrasses, polychaetes, shrimp, crabs, gastropods, and holothurians. It attains a maximum length of 70 cm and weight of 5.5 kg.

Venomous organs: It is a venomous and traumatogenic species. Its venomous organ consists of its prominent dorsal fin and two pectoral fins which have strong and sharply pointed, mucous-covered spines with serrations on the anterior (front) edges. This species has caused human envenomation. Its single dorsal spine and two pectoral spines are responsible for the venom delivery. The marine catfish envenomation has been reported to be severe than that of freshwater catfish (Dorooshi, 2012).

Ariopsis seemanni (Günther, 1864) (= *Arius jordani; Arius seemanni*)

Image credit: Xocolatl, Wikipedia. https://creativecommons.org/licenses/by-sa/3.0/

Common name (s): Tete sea catfish and Colombian shark catfish.

Global distribution: Widely distributed in the Pacific coastline of north-eastern tip of South America and Central America.

Ecology: This marine, freshwater, brackish, and demersal species dwells normally in river mouth.

Biology:

Description: Body of this species is moderately elongated and eyes are large. Head shield is granular. Gill opening is wide; gill rakers on first arch are 5–8 + 9–15; and there are no rakers on posterior face of first arch. Lips are thick and ruffled. Teeth are in four patches on palate. Pectoral spine is strongly serrated. Body coloration is dark blue to greenish brown on back; sides are greyish or with greenish-bronzy sheen; and white on lower surface. Fins are dusky and pelvic fins with wide white border. It attains a maximum length of 35 cm.

Food and feeding: While juveniles will take bloodworm, larger fish consume mainly earthworms, mussel, prawn, etc.

Reproduction: It is a paternal mouth brooder. It spawns in the ocean and male collects the eggs in its mouth. After the incubation, males deposit the fry in rivers where they spend their early life, before going back to the coastal waters.

Venomology: The gland which produces the venom is present at the base of the dorsal spine in this species. This venom may cause a lot of pain. (https://www.tankfacts.com/fish/freshwater/freshwater-shark/columbian-shark_412). It is also traumatogenic (https://www.fishbase.se/summary/13480).

Bagre bagre (Linnaeus, 1766)

Image credit: Para. José Milton Barbosa (Reproduced with permission).

Common name (s): Coco sea catfish.

Global distribution: Tropical; South America: Colombia to Amazon River.

Ecology: This marine, brackish, and demersal species has a depth range of 1–50 m Though it is chiefly marine, it is commonly seen around river mouths and also enters river estuaries.

Biology: Its head is arched and little depressed; and snout is rounded. There are two pairs of barbels on rostral region; one pair is on lower jaw and one pair is on posterior end of maxilla; and end of maxilla barbel is beyond anal fin origin. Body is silver gray or blue gray dorsally, and silver white ventrally. A large black blotch is seen on front of anal fin. Second dorsal adipose fin is well developed. Tail fin is deeply forked. It attains a maximum length of about 60 cm and its maximum published weight is 101 g. It consumes mainly invertebrates such as crustaceans and small fishes.

Venomous organs: It is a traumatogenic and venomous species (Anon., https://www.fishbase.in/summary/958; Haddad Jr., 2020). Spines of its dorsal and pectoral fins are with long flat filaments.

Bagre marinus (Mitchill, 1815)

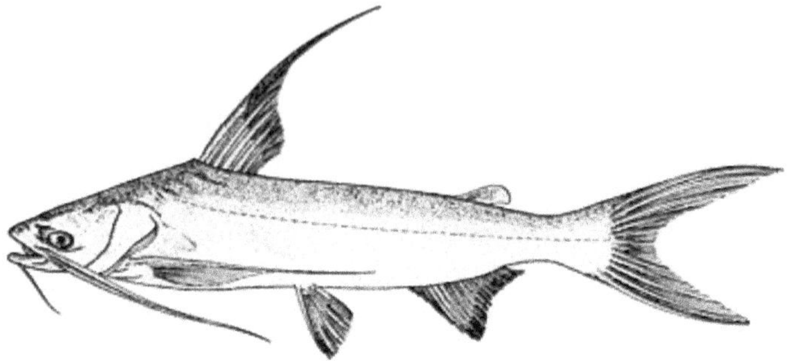

Source: Public domain

Common name (s): Gafftopsail sea catfish, and bigmouth catfish.

Global distribution: Subtropical; Western Atlantic: coast of Gulf of Mexico, Cuba, Caribbean, and South America.

Ecology: This demersal species which is common around bridges and piers and over hard substrates is mainly marine (neritic, intertidal, and coastal/supratidal) but enters brackish estuaries and mangroves with relatively high salinities. Its depth range is 0–50 m.

Biology: Body of this species is elongated and head is slightly depressed, with the mouth inferior. Its scaleless body is bluish above and white below. Among its three pairs of barbels, one pair is on the maxilla; and the other two pairs set under the chin. Maxillary pair of barbels are elongated reaching nearly to the ventral fins. Adipose fin is tipped in black, while the remainder of the fins are pale to dusky in color. Females have larger pelvic fins than males. It attains a maximum length of 70 cm and its main food items include crustaceans and small fishes.

Venomous organs: The potential venom delivery system which is the bony spine of this species is Y shaped. Paired anterior venom glands are present (Wright, 2012). McKinstry (1993) reported severe pain and bleeding in humans due to its sting. A serrated erectile spine present in its dorsal and pectoral fins is venomous (Anon., https://www.fishbase.se/summary/Bagre-marinus). Its long poisonous spines caused painful wounds (Anon., https://en.wikipedia.org/wiki/Gafftopsail_catfish).

Bagre panamensis (Gill, 1863)

Image credit: Robertson, Ross (Reproduced with permission).

Common name (s): Chilhuil sea catfish.

Global distribution: Subtropical; Eastern Pacific: southern California, USA to Peru.

Ecology: This marine and brackish species usually inhabits muddy bottoms of inshore areas.

Biology: Body of this species is very robust. Head shield is with ridges and not with granules. Lips are absent except at corner of mouth. There are only two pairs of barbels; and top barbel is broad and ribbon-like. Palate teeth are in four patches and are fused to form narrow, continuous band. Dorsal spine is without filament. Pectoral spine is with a long flattened filament reaching origin of anal fin. It reaches a maximum length of 51 cm.

Venomous organs: Spine present in front of its dorsal and pectoral fins is venomous and it can cause a painful wound (Anon., https://www.fishbase.se/summary/3056).

Batrachocephalus mino (Hamilton, 1822)

Source: Public domain

Common name (s): Beardless sea catfish.

Global distribution: Tropical; Indo-West Pacific: South and southeast Asian countries such as India, Sri Lanka, Malaysia, Indonesia, Myanmar, Thailand, Pakistan, and Bangladesh.

Ecology: This demersal species inhabiting coastal waters, estuaries, and tidal rivers.

Biology: In this species, maxillary barbel is absent but is with two pairs of mental barbels. Base of adipose fin is moderately long which is about half the length of anal-fin base. Lateral line is bifurcated at caudal region and is reaching base of caudal-fin's upper and lower lobes. It attains a maximum length of 25 cm. It prefers invertebrates and small fishes in its diet.

Venomous organs: Its potential venom delivery system is the bony spine which is Y shaped (Wright, 2012).

Paired anterior venom glands are present (Wright, 2012).

Carlarius heudeolotii (Valenciennes, 1840) (= *Arius heudeolotii*)

Common name (s): Smoothmouth sea catfish.

Global distribution: Eastern Atlantic: Nigeria.

Ecology: This marine, freshwater, brackish, and demersal species is commonly seen in the shallow continental shelf waters.

Biology

Description: Body of this species is elongated and rounded. Head is broad and is slightly flattened above; snout is rounded; and mouth is inferior. Its dorsal fin which is short has a long and bony serrated spine. Adipose fin is well developed; and caudal fin is forked, with long and pointed lobes. Body coloration is olive-brown above, silvery on sides, and white in belly. It attains a maximum length of 83 cm.

Food and feeding: It feeds mainly on benthic invertebrate animals of the muddy bottoms. However, it is capable of capturing its prey in open water.

Reproduction: Female of this species bears large eggs which seem to be incubated and developed in the male's buccal cavity.

Venomology: This venomous species has dorsal and pectoral fins which are with a strong serrated, erectile spine. The wound inflicted by this species is extremely painful (https://www.fishbase.de/summary/Carlarius-heudelotii.html; https://www.fishbase.se/summary/4928).

Carlarius latiscutatus Günther, 1864 (=*Arius latiscutatus*)

Common name (s): Rough-head sea catfish.

Global distribution: Tropical; Eastern Atlantic: from Senegal to Angola.

Ecology: This coastal, brackish and demersal species inhabits up to a depth of 70 m. It is also frequently found in estuaries and sometimes entering freshwater and lagoons.

Biology: Body of this species is elongated and rounded. Head is more rounded and is only slightly flattened above; snout is rounded and mouth is inferior. Its maxillary barbel does not reach beyond the pectoral fin base. Adipose fin is well developed. It attains a maximum length of 85 cm and its diet consists of mainly benthic invertebrates, zooplankton, and fish.

Venomous organs: It is a venomous species (Anon., https://www.fishbase.in/summary/4929) and its dorsal and pectoral fins which are with a strong serrated, erectile spine may harm humans.

Cathorops spixii (Agassiz, 1829)

Image credit: Kolding, Jeppe (Reproduced with permission).

Common name (s): Yellow catfish, Madamango sea catfish.

Global distribution: Tropical; Western Atlantic: Common in the estuaries of Colombia to Brazil; and Atlantic and Caribbean rivers.

Ecology: This demersal species inhabits all marine, brackish and freshwater areas at a depth range of 1–50 m. It may also be seen in hypersaline waters.

Biology: Three pairs of barbels are seen on its rostral region. Among them, one pair is seen on the posterior end of maxilla and two pairs on its lower jaw. Body coloration is dark brown or black dorsally, and pale yellow ventrally. Fin membranes are brown, with small black spots. Head is widely flattened from above and mouth is inferior. Nostril is rounded, Adult fishes feed mainly on benthic invertebrates and fishes; and Juveniles on copepods, amphipods, and isopods. It reaches a maximum length of about 30 cm.

Venomous organs: The skin mucus and sting venom of this species exhibit harmful effects. While the peptide fractions caused venular stasis, the protein fractions produced an inflammatory process in post-capillary venules. Further, the protein WAP65 isolated from its sting venom also showed inflammatory action inducing an increase in the number of leukocytes rolling and adhering to the endothelium (Ramos et al., 2012; Haddad Jr. et al., 2014).

Cephalocassis borneensis (Bleeker, 1851) (= *Hemipimelodus borneensis*)

Image not available.

Common name (s): None.

Global distribution: Tropical; Western Central Pacific: Sundaland and Indochina; Thailand to Indonesia.

Ecology: This demersal species is potamodromous occurring in large lowland rivers. It may also thrive in brackish waters.

Biology: Body shape of this species is laterally elongated. Head is about as deep as wide with rounded snout but without teeth on palate. Caudal fin is forked. Its diets consists mainly of bivalve mollusks, crustaceans, aquatic macrophytes, and detritus. Adipose fin is present. It attains a maximum length of 30 cm.

Venomous organs: The potential venom delivery system which is the bony spine of this species is Y shaped, small, and its venom glands are posterior (Wright, 2012).

Cochlefelis danielsi (Regan, 1908)

Common name (s): Daniel's catfish.

Global distribution: Tropical; Asia and Oceania: New Guinea.

Ecology: This demersal species occurs in turbid rivers and in swampy lagoons and backwaters. It may also thrive in brackish waters including mangrove areas.

Biology: It feeds mainly on the crustaceans viz. *Caridina* sp. and *Macrobrachium* sp. and it attains a length of about 50 cm (Anon., https://en.wikipedia.org/wiki/"Caridina"\o"Caridina").

Venomous organs: The potential venom delivery system which is the bony spine of this species is Y shaped. Posterior glandular cells are visible (Wright, 2012).

Galeichthys feliceps (Valenciennes, 1840)

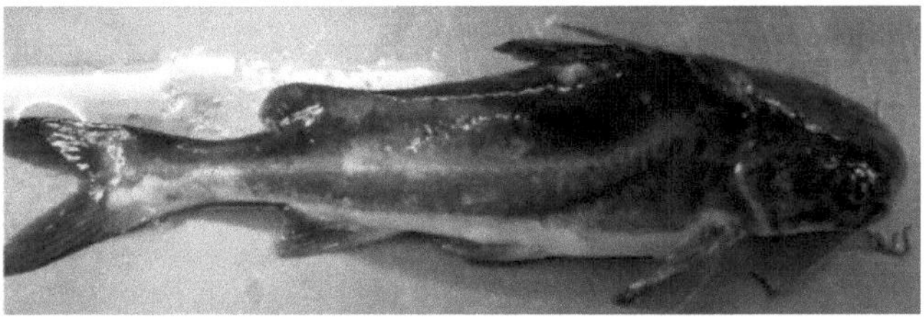

Image credit: Frederick Hermanus Van der Bank Wikipedia. Copyrighted-free use.

Common name (s): White barbel.

Global distribution: Subtropical; Southeast Atlantic: Namibia to East London, South Africa.

Ecology: This demersal and amphidromous species inhabits both marine, brackish water and riverine regions up to a depth of 120 m. It is more common in muddy bottoms of turbid coastal and estuarine waters. It may also form large shoals.

Biology: This species is distinguished by characters such as longer head; deeply forked tail fin with lean and sharp lobes; presence of 11–14 gill-rakers on the anterior face of the first gill arch; thin and long maxillary barbel; darkened upper two-thirds of body; stark white belly; and 10–12 pectoral fin rays. It consumes mainly crayfish, crabs, and small fish. It attains a maximum length size of about 60 cm in length.

Venomous organs: Its venomous spines may cause wounds which should be treated immediately (Anon., https://www.fishbase.de/summary/288).

Genidens (Cuvier, 1829)

Source: Public domain

Common name (s): Guri sea catfish, snake catfish.

Global distribution: Subtropical Southwest Atlantic: eastern coast of South America.

Ecology: It is commonly found in estuaries and coastal lagoons. However, it is known to spend its entire life cycle in the estuaries. It is also displaying high tolerance to eutrophic conditions.

Biology: In this species, the accessory plates of its teeth are not fixed in the palatal region but are linked to fleshy projections. It attains a maximum length of 42.5 cm.

Venomous organs: The serrated bony sting of its dorsal fin and each of its pectoral fins is traumatogenic. The venom-bearing glandular tissues are found diffused under the epithelium which covers the stings. Its venom components include biogenic fractions, the functions of which are very much similar to that of the stingray venom (Haddad Jr et al., 2008). Junqueira et al. (2006) reported that the sting of this species was found contaminated with 13 different strains of Enterobacteriaceae with dominant species such as *Klebsiella pneumoniae, Enterobacter sp., Escherichia coli, Serratia marcescens, Serratia sp., and Proteus mirabilis.* This sort of contamination in these fish stings lead to significative acute secondary infections in humans.

Neoarius graeffei (Kner and Steindachner, 1867)

Image credit: Richard Fuller, Wikimedia Commons. https://creativecommons.org/publicdomain/zero/1.0/deed.en

Common name (s): Salmon catfish or blue salmon catfish.

Global distribution: Tropical Western Pacific: Southern New Guinea and Northern Australia.

Ecology: This demersal, anadromous species dwells in a variety of biotopes such as lagoons, rivers, estuaries, and coastal waters.

Biology: Back of all gill arches are with raker like processes. Palatel teeth are villiform in transverse series of four oval patches; inner (vomerine) patches are smaller than outer (palatine) patches; and patches may fuse to form two patches or one large patch in larger specimens. Maxillary barbels are moderately long and are extending to edge of opercle. Head is ovate and snout is rounded or slightly truncated. Body coloration is dark brown or dark blue above and yellowish, cream or white ventrally. Fins are bluish brown. It feeds mainly on insect larvae, aquatic plants, prawns, crayfish, molluscs, small fishes, and bottom detritus. It attains a maximum length of 60 cm.

Venomous organs: It is a traumatogenic (Anon., https://www.fishbase.in/summary/22590) and venomous species (Anon., https://fishesofaustralia.net.au/home/species/2141).

Netuma bilineata (Valenciennes, 1840)

Image credit: Gloerfelt-Tarp, T. (Reproduced with permission).

Common name (s): Bronze catfish, Two-line Sea Catfish.

Global distribution: Tropical Indo-West Pacific: Northern Australia, Queensland, and Japan; and Persian Gulf to Indo-Malayan area.

Ecology: This demersal species inhabits all habitats viz. freshwater, estuary, and coastal waters. It is found mostly on mudflats.

Biology: Its dorsomedian head groove is bordered by a fairly raised frontals in the shape of a "V" at its posterior end. Its first two gill arches lack gill rakers. Tail fin lobes are slender and tapered. Adipose fin is short-based and located posteriorly. It is mostly opportunistic and carnivorous feeding mainly on crustaceans, fish, prawns, sea urchins, and detritus. It attains a maximum length of 90 cm.

Venomous organs: This traumatogenic fish has sharp dorsal and pectoral spines which are said to inflict painful injuries (Anon., https://www.fishbase.se/summary/Netuma-bilineata.html). Junqueira et al. (2006) reported that the toxic extract from the skin of this species caused death in the tested animals when intravenously injected.

Netuma thalassina (Rüppell, 1837) (= *Arius thalassinus*)

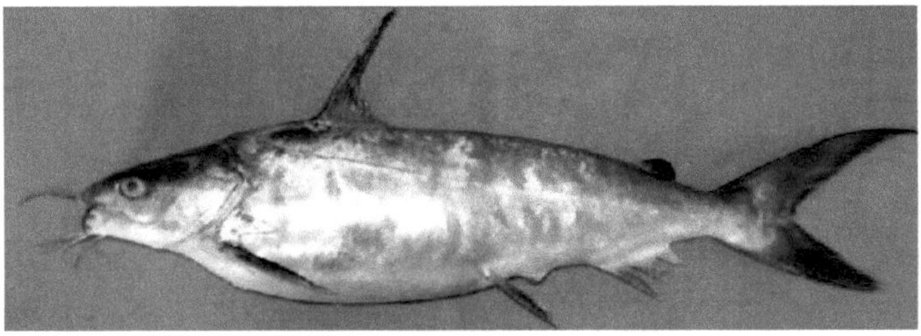

Image credit: Hamid Badar Osmany, Wikimedia. https://creativecommons.org/licenses/by/3.0/

Common name (s): Arabian Gulf catfish; giant marine-catfish, and giant salmon catfish.

Global distribution: Subtropical; Indo-West Pacific: Red Sea and the northwestern Indian Ocean; and also from Australia, Polynesia, and Japan.

Ecology: This demersal, amphidromous and marine species is often found in estuaries, but it rarely enters freshwater. Its depth range is 10–195 m.

Biology: In this species, snout is of varying shape. While in males it is acute and pointed, it is rounded in females. Three pairs of barbels are seen around mouth. Maxillary pair is almost reaching pectoral fin base. Head shield is weakly granulated and rugose. First dorsal and pectoral fins each with a strong venomous spine. Males of this species are dark-red-brown to bluish-grey above and is densely pigmented below. Entire body is with a bronze luster. Numerous narrow, green cross bands may occasionally be seen in fresh specimens. Distal part of dorsal adipose, anal and caudal fins are dark. Females are uniformly dark or grey. Diet of this species consists of mainly worms, prawns, stomatopods and crabs; and cephalopods, sea cucumbers, and finfish. It attains a maximum length of 185 cm and weight of 1 kg.

Venomous organs and Envenomation: It is a Traumatogenic (Anon., https://www.fishbase.de/summary/10220) and venomous species. White and Meier (2017) reported that its crinotoxin (toxic epidermal secretions) plays a significant role in the spinous envenomation process. Blomkalns and Otten (1999) also reported that this species releases the crinotoxin when the catfish is excited or threatened. If exposed to open skin, these toxins cause similar symptoms of spine toxicity such as throbbing pain, tissue necrosis, and muscle fasciculation.

Osteogeniosus militaris (Linnaeus, 1758)

Image credit: *Staticd*, Wikimedia. https://creativecommons.org/licenses/by-sa/3.0/

Common name (s): Soldier catfish.

Global distribution: Tropical; Indo-West Pacific: Bangladesh, Pakistan, India, Malaysia, Singapore, Myanmar, Malacca, and Indonesia.

Ecology: This benthic and potamodromous species dwells in coastal waters and estuaries.

Biology: Body coloration of this species may be yellowish or slate blue. It has a single pair of maxillary barbels which are stiff and sharp. It produces a thick mucus all over its body especially around the head. It feeds mainly on invertebrates and small fishes. It attains a maximum length of 42 cm.

Venomous organs: It is traumatogenic species (Anon., https://www.fishbase.se/summary/Osteogeneiosus-militaris.html) and its spines are venomous (Fish and Cobb, 1954).

Sciades herzbergii (Bloch, 1794)

Image credit: Clara Costa D'Elia, Wikimedia. https://creativecommons.org/licenses/by/4.0/

Common name (s): Pemecou sea catfish.

Global distribution: Tropical; South America: Caribbean and Atlantic rivers and estuaries from Colombia to Brazil.

Ecology: This demersal species occurs in turbid estuaries, mangrove-lined lagoons and lower parts of rivers. It prefers shallow soft bottoms.

Biology: It has a characteristic furrow which is partly covered by a flap of skin. Snout is long and rounded; and mouth is slightly inferior. There are usually 21–23 gill rakers on second arch. Palatine teeth patch is always with a U-form. Body coloration is grey color to dark brown above and whitish below. All fins are with a touch of red/pink. It feeds on a e variety of bottom-living and pelagic organisms including that of the roots of mangroves. As a medium sized species, it attains a maximum length of 94 cm and weight of 1.5 kg.

Venomous organs: it is a traumatogenic species (Anon., https://www.fishbase.se/summary/951). Macêdo et al. (2017) reported that this species collected from the Paraiba estuary (Brasil) is venomous. However, its toxin is not known.

4.3 NAKED CATFISHES OR BAGRID CATFISHES (BAGRIDAE)

Mystus cavasius (Hamilton, 1822)

Image credit: Dr. Mohammad Moazzam Khan (Reproduced with permission).

Common name (s): Gangetic mystus.

Global distribution: Tropical Asia: Indian subcontinent.

Ecology: It is a demersal species occurring in tidal rivers and lakes; and also in canals, ditches, ponds, and beels. This amphidromous species may tolerate brackish conditions also.

Biology: Body of this species is elongate and compressed. Head is conical. Barbels of upper jaw extend posteriorly beyond the base of the tail fin.

Dorsal spine is weak and is very slightly serrated. Body coloration is grayish with a well-defined mid-lateral longitudinal stripe. A dark spot with a white or pale area is seen along its ventral margin. Dorsal, adipose and caudal fins are with melanophore shades. It attains a maximum length of 40 cm.

Venomous organs: Its pectoral spine can cause painful wounds (Anon., https://www.fishbase.in/country/CountrySpeciesSummary.php?c_code=104&id=5140).

Mystus gulio (Hamilton, 1822)

Image credit: Hammad Eusuf Hasan* (Reproduced with permission).

Common name (s): Long whiskers catfish.

Global distribution: Tropical; Asia: India to Indonesia; Viet Nam and Pakistan.

Ecology: Freshwater; brackish; demersal. It is primarily a brackish water fish entering and living in fresh water. In freshwater, adults occur mainly in rivers and streams with mud or clay substrates. It may form schools. It is anadromous and diurnal.

Biology: Body of this species is elongated and compressed. Head is depressed and its upper surface is rough and granulated. Snout is broad and slightly depressed; mouth is terminal; upper jaw is somewhat longer; and teeth are villiform in bands on jaws. There are four pairs of barbels and maxillary ones extend posteriorly to end of pelvic fins. Dorsal spine is strong and is serrated on its inner edge. Adipose fin is small and is inserted well behind rayed dorsal fin. Pectoral fin does not reach the ventral and the length of its spine more or less the same with that of its head. Caudal fin is forked, with longer upper lobe and rounded lower one. Body coloration is bluish-brown on the head and back and becoming dull white beneath. Fins are normally black on their outer halves; and maxillary barbels are black, whitish, or white tipped. Male fish has an elongate genital papilla in front of its anal fin. Its food consists mainly of crustaceans and insects. It reaches a maximum length of 46 cm.

Venomous organs and envenomation: The crude spine sheath extract venoms of this species exhibited analgesic properties on male albino mice. Further, this extract possesses hemolytic activity by inducing spontaneous hemolysis of chicken blood (Baskaradoss and Kumar, 2018).

4.4 THORNY CATFISHES (DORADIDAE)

Doras carinatus (Linnaeus, 1766)

Image credit: Holm, Erling, (Reproduced with permission).

Common name (s): Thorny catfish.

Global distribution: Tropical South America.

Ecology: This demersal and euryhaline species inhabits both brackish estuaries and freshwaters.

Biology: Body of this species is elongated and without scales. However, a spiny bony plaque is present and this runs laterally from the end of the caudal fin up to the beginning of the dorsal fin. Head is almost sharp and is laterally compressed. Eyes are high and are located at the rear end of the head. Its inferior mouth has a pair of maxillary barbels which branch into 13 branches and 4 fairly short mandibular barbels. Adipose fin is short. It reaches a maximum length of about 30 cm.

Venomous organs: This species is armed with formidable spines which are capable of inflicting dangerous wounds (Fleming, 2005).

4.5 EELTAIL CATFISHES (PLOTOSIDAE)

Cnidoglanis macrocephalus (Valenciennes, 1840)
(= *Cnidoglanis* megastoma)

Image credit: Sylke Rohrlach Wikimedia Commons. https://creativecommons.org/licenses/by-sa/3.0/

Common name (s): Estuary catfish, estuary cobbler, Swan River catfish or South Australian catfish.

Global distribution: South and Western Australia.

Ecology: This marine and brackish water species dwells in sandy inlets and shallow bays nearer to river mouths. It is demersal and is mostly found over seaweeds, sand, and rocks. During day time, these fishes are often found on ledges or in crevices nearer to the river banks. They are nocturnal in their feeding habit.

Biology: It has a characteristic catfish body in front and eel body behind. It is a large mottled brownish-yellow fish with robust dorsal and pectoral-fin spines. There are five pairs barbels around its small mouth. Its fifth pair of barbels is seen above the snout. It has a large flattened head and its body tapers to a pointed tail. Dorsal, anal and tail fins are continuous. These are opportunistic feeders, and are feeding mainly on crustaceans, molluscs, polychaete worms, algae, and bottom debris. It attains a maximum length of 91 cm and weight of 2.5 kg.

Venomous organs: These fishes possess strong dorsal and pectoral-fin spines which are capable of inflicting very painful wounds. Puncture wounds due to this fish can be very painful (Anon., http://fishesofaustralia.net.au/home/species/2758). Dart (2004) reported that this species has caused human envenomations in Indo-Pacific region and Australia.

Paraplotosus albilabris (Valenciennes, 1840)

Image credit: Helen Sykes (Reproduced with permission).

Common name (s): Whitelipped eel catfish.

Global distribution: Tropical; Indo-West Pacific: Indo-Australian Archipelago.

Ecology: This demersal and reef-associated species dwells both in coastal and brackish waters at a depth range of only 1–12 m. It is often found in areas with mixed sand, weed, and corals. Adults may be solitarily or in small groups and they are nocturnal.

Biology: This species has shorter dorsal fin and shorter nasal barbels; rounded pectoral fin; and fewer upper procurrent caudal fin rays and gill rakers. Body coloration is highly variable and is ranging mainly from yellowish brown to dark brown. Fins are brown to blackish. It normally feeds on crustaceans and gastropods. It attains a maximum length of 1.3 m.

Venomous organs: The venom delivery system which is the bony spine of this species is Y shaped. Large, paired, dorsal, and ventral venom glands are present in the spine (Wright, 2012).

Plotosus canius (Hamilton, 1822)

Source: Public domain

Common name (s): Gray eel-catfish.

Global distribution: Tropical; Indo-West Pacific: India and off Sri Lanka; Bangladesh and Myanmar; Indo-Australian Archipelago and Philippines; Papua New Guinea.

Ecology: Adults are found mostly in estuaries and lagoons, and sometimes it may also be found in the lower parts of rivers, brackish water, and in coastal seas.

Biology: Body of this species is elongated, anteriorly depressed, and posteriorly compressed. Mouth is transverse with a fairly longer upper lip. Snout is rounded. Dorsal and pectoral spines are serrated on both edges. Second dorsal and anal fins are confluent with the caudal. There are four pairs of barbels and among them, maxillary is long and nasal and mandibular barbels reach beyond the gill-opening. It is a plain dusky-brown species with a black dorsal fin tip. It shows banded pattern at night. It feeds on crustaceans, mollusks, and fishes and it has a maximum length of 150 cm.

Venomous organs and envenomation: The bony spine of this species which is the potential venom delivery system is Y shaped. Large, paired, hemispherical venom glands are seen in the venomous spines associated with anterior fins (Wright, 2012; Anon., https://www.fishbase.se/summary/Plotosus-canius). Auddy et al. (1995) reported that the lethal protein toxin of this species was cardiotoxic with neuromuscular blocking activity. This toxin has been reported to produce cardiac arrest on the experimental hearts of toad and guinea pig.

Plotosus japonicus Tetsuo Yoshino and Hirokazu Kishimoto

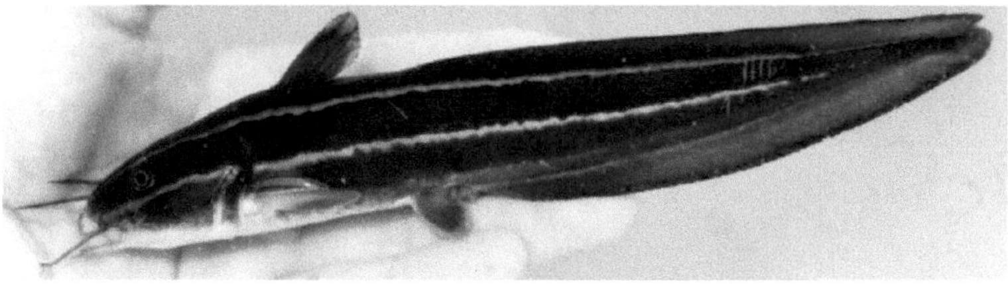

Image credit: ffish.asia. https://creativecommons.org/licenses/by/4.0/

Common name (s): Eeltail Catfish.

Global distribution: Temperate; Northwest Pacific: Japan.

Ecology: It is a benthopelagic species.

Biology

Description: Body of this species is elongated and compressed with a tapering tail. Head is depressed and broad. Snout margin is rounded when viewed from above. Mouth is subterminal; and lips are thick with laminated folds. There are four pairs of barbels and each barbel is somewhat broadened and flattened. While the nasal barbel is extending slightly over posterior margin of eye, maxillary barbel is extending well beyond posterior margin of eye. Its two mandibular barbels are extending to level of pre-operculum. Jaws are with conical teeth. Dorsal fin is single and are located posterior to head. Pectoral fin is rounded. Pelvic fin is with rounded distal margin. Adipose fin is absent. Caudal fin is lanceolate. Anal and caudal fins are confluent, forming a continuous median fin along more than posterior half of body dorsally and ventrally. In fresh specimens, head and body are dark brown, becoming cream or white ventrally. There are two to three narrow pale-yellow longitudinal stripes on body, of which two are extending onto head. Barbels are dark brown and are gradually becoming paler distally. Anal and caudal fins are hyaline with dark brown distal margins. In alcohol-preserved fishes, coloration of body and head is brown above and paler below. It attains a maximum length of 18 cm. (Yoshino and Kishimoto, 2008).

Venomous organs: It is a venomous fish (Clark et al., 2011). Its venomous organs are its dorsal fin with a sharp spine serrated with fine basally directed teeth on both edges; and its pectoral fin is with a sharp spine serrated with fine antrorse teeth on both edges.

Plotosus limbatus (Valenciennes, 1840)

Image credit: Mr. & Mrs. Richard; Dennis Polack (Reproduced with permission).

Common name (s): Darkfin eel catfish.

Global distribution: Tropical Western Indian Ocean: From the coasts of Chennai and Malabar (India) to the east coast of Africa.

Ecology: This demersal and amphidromous species occurs in estuaries and all along open coasts.

Biology: Head of this species is moderately large and its profile is arched from the tip of snout to the origin of dorsal fin. There are four pairs of barbels. Nasal barbels extend to posterior borders of eyes and maxillary barbels extend from opercular margin to pectoral spine origin. Anterior nostrils are located above edge of upper lip. Eyes are moderate to large and teeth of upper jaw are pointed and are in two rectangular patches of two to four rows each. About 20 gill rakers are seen on anterior edge of first arch. Body coloration in preserved specimens is reddish brows to dark brown. Fins are often blackish brown, usually with a black border. It feeds on crustaceans, mollusks, and fishes and attains a maximum length of 50 cm.

Venomous organs: Dorsal and pelvic fins of this species are venomous (Anon., https://www.fishbase.se/summary/Plotosus-limbatus.html).

Plotosus lineatus (Thunberg, 1787)

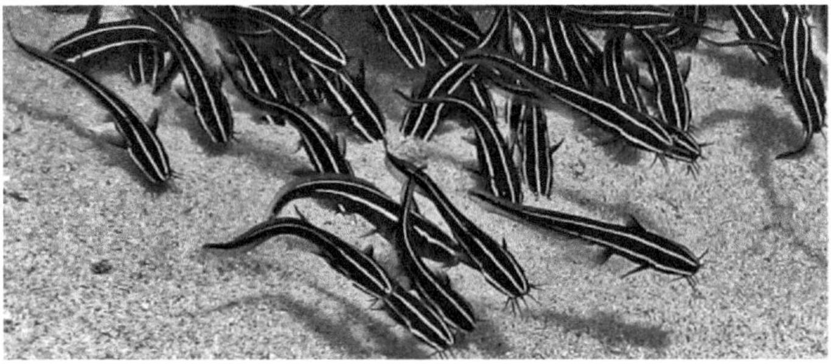

Image credit: Hectonichus, Wikipedia. https://creativecommons.org/licenses/by-sa/3.0/

Common name (s): Striped eel catfish, Coral Catfish.

Global distribution: Tropical; Indo-Pacific: East Africa, Madagascar, Japan, southern Korea, and Ogasawara Islands.

Ecology: This reef-associated and amphidromous species is commonly found in estuaries, tide pools, coral reefs, and open coastal waters at a depth range of 1.60 m. This species may form schools.

Biology: Body of this species is elongated and torpedo shaped, and compressed. Head is large. Dorsal and anal fins are continuous with caudal fin. There are four pairs of short mouth barbels. Body coloration is dark brown to blackish above and is becoming fawn toward the belly. Fins are brown and the combined fin is often dark edged. A single highly venomous serrate spine is seen at the start of the first dorsal fin and in each of the pectoral fins. It may attain a total length of 32 cm. Its food items include worms, crustaceans, mollusks, and fish. Its feeding behavior is unique. During feeding, it acts as a giant rotary machine, taking turns with its snout in the substrate to look for food and hanging up top to keep an eye out for its predators.

Venomous organs and envenomation: The concealed spines of the pectoral and first dorsal fins deliver a highly toxic venom when touched and this may be fatal to humans (Anon., https://www.leisurepro.com/blog/explore-the-blue/march-plotosus-lineatus/). Blomkalns and Otten (1999) also reported that this species also releases the crinotoxin when the catfish is excited or threatened. If exposed to open skin, these toxins cause similar symptoms of spine toxicity such as throbbing pain, tissue necrosis, and muscle fasciculation.

Plotosus nkunga (Gomon and Taylor, 1982)

Image not available.

Common name (s): Stinging eel catfish.

Global distribution: Tropical Western Indian Ocean: South Africa; and Zanzibar and Tanzania.

Ecology: This demersal species is normally found in estuaries and along open coasts. It may however, enter freshwaters.

Biology: Head of this species is moderately large and its profile is slightly arched from tip of snout to dorsal fin origin. There are four pairs of barbels. Nasal and maxillary barbels extend at most to hind border of eyes and anterior nostrils are located on end of snout above edge of upper lip. Eyes are moderate to large. Teeth in upper jaw are pointed and are in two rectangular or triangular patches of two to four rows each. There are 16–21 gìll rakers on anterior edge of first arch. In preserved specimens, body is greyish to reddish brown and is fading to a lighter brown ventrally. Two faint, pale longitudinal stripes are sometimes present on body, but they are not extending onto head. It attains a maximum length of 51 cm.

Venomous organs: First dorsal fin is with one spine which is venomous (Anon., http://www.fao.org/3/ad468e/ad468eJG.pdf).

Tandanus bostocki (Whitley, 1944)

Image credit: Healthy Rivers, Department of Water and Environmental Regulation, Western Australia (Reproduced with permission).

Common name (s): Freshwater cobbler.

Global distribution: Oceania: endemic to Australia.

Ecology: This demersal species lives in slow-flowing streams, ponds and reservoirs. It may also occur in brackish waters, isolated pools of river-beds, and lakes. Underwater cavities of river banks and root mounds may are utilized as shelter by this species.

Biology: It has an eel-like body with continuous second dorsal, anal and caudal fins. Four pairs of barbels (whiskers) are seen surrounding the mouth. Body coloration is dark brown. In clear waters clear mottling along the entire body and head could be seen. It may attain a maximum length of 50 cm.

Venomous organs: The dorsal and pectoral fins are armed with venomous spines that may inflict a very painful sting. (Anon., https://fishesofaustralia.net.au/home/species/3311).

4.6 GLASS CATFISHES (SCHILBEIDAE)

Eutropiichthys vacha (Hamilton, 1822)

Source: Public domain

Common name (s): Batchwa vacha.

Global distribution: Tropical; Asia: Pakistan, India, Bangladesh, Nepal, Myanmar, and Thailand.

Ecology: This pelagic, potamodromous species is found in rivers, canals, and tidal waters. It may enter brackish waters also.

Biology: Body of this species is elongated and is laterally compressed. Its dorsal and ventral profile are more or less convex. Upper jaw is slightly longer than lower. It has four pairs of barbels viz. one pair of nasal, one pair of maxillary, and two pairs of mandibular. Nasal barbels often reach just behind the head while maxillary barbels reach the base of pectoral fin; and mandibular barbels are normally smaller and they do not extend beyond the head. Body coloration is silvery with grayish back. Dorsal spine is serrated posteriorly and pectoral fin is serrated internally. Adipose fin is present and caudal fin is deeply forked. Pectoral and anal fins have reddish margin. Lateral line is complete. It is known to feed mainly insects and small fish. Its maximum length and weight is 38 cm and 1.4 kg, respectively.

Venomous organs: The potential venom delivery system which is the bony spine of this species is Y shaped. Spine has no visible venom glands (Wright, 2012).

Silonia silondia (Hamilton, 1822)

Source: Public domain

Common name (s): Silond catfish.

Global distribution: Tropical; Asia: Pakistan, India, Bangladesh, Nepal and Myanmar.

Ecology: This demersal and amphidromous species inhabits mainly rivers. For the purpose of breeding, adult fishes migrate to large rivers from their estuarine habitats during monsoon season. This species has been reported to get stranded in small pools as and when the water level recedes.

Biology: Body of this species is elongated and is deeply compressed. Mouth is terminal with a little longer lower jaw. Snout is broad and rounded. Eyes are with narrow adipose lids. Barbels are in two pairs. Dorsal spine is generally weak than pectoral spine and both spines are however, finely serrated posteriorly. Body coloration is yellowish-green on back, and silvery purple on flanks and abdomen. Golden tinge is present on both sides of head. Caudal, anal, and pelvic bases are yellowish. It is carnivorous and voracious feeder on fish, organic matters, etc.; it is known to attain a maximum length of 100 cm and weight of 4 kg.

Venomous organs: The potential venom delivery system which is the bony spine of this species is Y shaped. Spine has visible glandular cells (Wright, 2012).

4.7 SHEAT CATFISHES (SILURIDAE)

Ompok bimaculatus (Bloch, 1794)

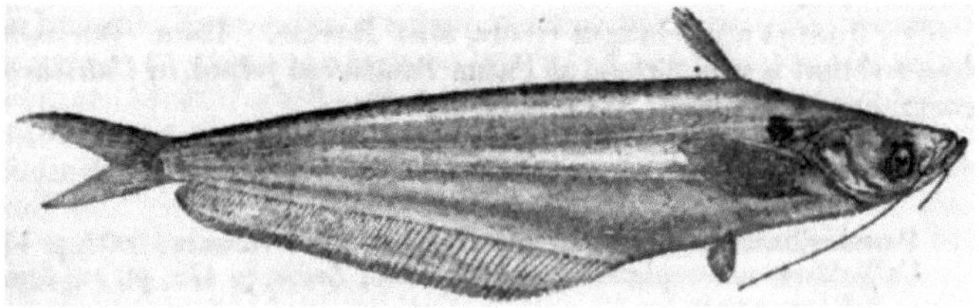

Source: Public domain

Common name (s): Butter catfish, sheatfish.

Global distribution: Tropical; Asia: Indian subcontinent and Myanmar.

Ecology: Though this demersal species dwells both in freshwater and brackish water estuaries, adults are mostly found in quiet, shallow muddy water; canals, beels, and inundated fields; sandy streams, and rivers and tanks.

Biology: In this species, its dorsal and pectoral fins are small and anal fin is long. Caudal fin is forked. It has two pairs of barbels viz. maxillary and mandibular pairs. Its smaller eyes are covered by skin. Its vomerine teeth are in in two patches. Coloration of body is silvery. A distinct, rounded black blotch is seen above and behind the base of the pectoral fin. One spot is located at the base of the caudal peduncle. It consumes mainly small fish and vegetable matter. It attains a maximum length of 45 cm.

Venomous organs: The venom delivery system which is the bony spine of this species is Y shaped. Posterior venom glands are visible in the spine (Wright, 2012).

Silurus glanis (Linnaeus, 1758)

Image credit: Andreas Hartl (Reproduced with permission).

Common name: Sheatfish, wels catfish, European catfish.

Global distribution: Temperate Europe and Asia.

Ecology: This benthopelagic and non-migratory species inhabits mainly in backwaters and well vegetated lakes; and large lakes and rivers at a depth range of 0–30 m. It is known to enter brackish water in the Baltic and Black Seas rarely. It is also found in deep waters of dams present in the lower reaches of rivers.

Biology: Its body is naked with a large, depressed head. Two pairs of mental barbels are seen. It has razor sharp teeth. Caudal fin is rounded or truncate and there is no adipose fin. Anal rays are almost touching caudal fin. It is a nocturnal predator, foraging animals near the bottom and in water column. It has a maximum length of 500 cm and weight of 306 kg.

Venomous organs: It has a dorsal fin with a spine and a paired pectoral fins with one spine in each. These spines are venomous. It is a very dangerous fish and an entire human body was found in the stomach of this species in Russia (Anon., https://whatdewhat.com/dangerous-freshwater-fish-can-kill/).

KEYWORDS

- **common name**
- **global distribution**
- **ecology**
- **biology**
- **venomous organs**

CHAPTER 5

TRAUMATOGENIC CATFISHES

ABSTRACT

The aspects of biology and ecology of traumatogenic catfishes belonging to three familiesof the Order Siluriformes viz. Ariidae, Cetosidae and Siluridae are dealt with in this chapter

There are several dangerous species of catfishes which are known to cause injuries to humans through their spines, stingers, and teeth. Some species on the other hand, are known to be more frequently associated with traumatic injuries. Such cases may not involve envenomation, but they may be associated with lacerations, severe bleeding and late microbial infections. This chapter deals with such species of catfishes inhabiting freshwater and marine environments.

5.1 ARIIDAE–ARIID CATFISH

Amissidens hainesi (Kailola, 2000)

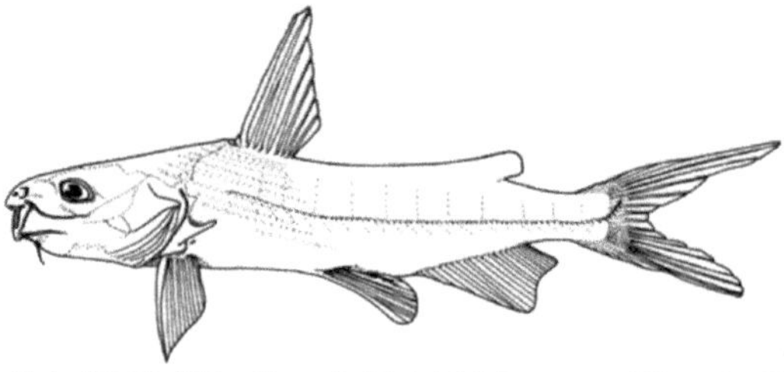

Image credit: P. Kailola 2000 / The Beagle Vol 16 Museums Victoria CC; https://creativecommons.org/licenses/by/3.0/au/

Common name (s): Ridged catfish.

Biology and Ecology of the Venomous Catfishes. Ramasamy Santhanam, PhD (Author)

Global distribution: Western Central Pacific: Northern Australia and Southern New Guinea.

Ecology: This benthic species is found mainly in the lower estuaries and coastal waters.

Biology

Description: Eyes of this species are large. Mouth is small and is almost quadrangular. Lips are fleshy and thin. Barbels are thin and short; and maxillary barbels extend beyond eye. Spines of fins are thin and long. Adipose fin which is with a short base is situated at the posterior end of the anal fin. A scalloped and tapered ventral fin pad is present in matured females. Dorsum of body is dark gray colored. This fish reaches a maximum length of 32 cm.

Food and feeding: It consumes aquatic insect larvae, gastropod mollusks, and crabs.

Reproduction: Not reported.

Traumatogenic structures: It is a traumatogenic species (https://www.fishbase.se/country/CountrySpeciesSummary.php?c_code=036&id=60007) and details of its traumatogenic structures are wanting.

Amphiarius phrigiatus (Valenciennes, 1840)

Image is not available.

Common name (s): Kukwari sea catfish.

Global distribution: Western Atlantic: From Venezuela to Brazil's Amazon River.

Ecology: This benthic species is seen in the shallow muddy bottoms of brackish estuaries and is known to enter freshwater.

Biology

Description: It reaches a total length of 35 cm.

Food and feeding: Not reported.

Reproduction: It is a mouth brooder and female lays her eggs on a sandy depression. Males guard the eggs until they are hatched.

Traumatogenic structures: It is a traumatogenic species (https://www.guppies.za.net/species.aspx?name=kukwari-sea-catfish-amphiariusphrygiatus&fid=28924) and details of its traumatogenic structures are wanting.

Ariopsis guatemalensis (Günther, 1864)

Common name (s): Blue sea catfish, wide hea sea catfish.

Global distribution: Pacific coast of Central America.

Ecology: This marine, brackish, and demersal species is also known to enter freshwater.

Biology

Description: This species has a moderately elongated body. Head is wide, long, flat; and head shield is rough. Eye is small. Gill rakers on first arch are 6–7 + 11–14. Adipose dorsal fin base is about 3/4 of first dorsal fin base. Body coloration is bluish black or charcoal on back; greenish violet on sides; and silvery white on lower surface. Fins are dusky to blackish. It has a maximum length of about 50 cm.

Food and feeding: It is a carnivorous species with a preference for goby fish. Crustaceans form the second most important food item and are composed of shrimps and crabs. It is also known to eat live and hard-bodied animals.

Reproduction: There is no breeding report.

Traumatogenic structures: This traumatogenic species (T- https://www.fishbase.se/summary/13471) has a pectoral spine which is very strong and is with prominent serrations on inner margin (https://biogeodb.stri.si.edu/sftep/en/thefishes/species/568).

Ariopsis seemanni (Günther, 1864)

Image credit: Xocolatl, Wikimedia. https://creativecommons.org/licenses/by-sa/3.0/

Common name (s): Tete sea catfish and Colombian shark catfish.

Global distribution: Central and South America; and Pacific rivers and estuaries.

Ecology: This demersal species dwells in marine, freshwater, and brackish water biotopes.

Biology

Description: Body of this species is moderately elongate and eyes are fairly large. Head shield is granular. Gill opening is wide. Lips are thick and ruffled. Body coloration is thick blue to greenish brown on dorsum; laterally, grayish or greenish; and belly is white. Fins are dusky; and pelvic fins are white bordered. It attains a maximum length of 44 cm.

Food and feeding: Juveniles feed on *bloodworm* while larger fish prefer earthworms, mussel, and prawns.

Reproduction: Sexually matured females are with fairly thick body than their male counterparts. Spawning takes place normally in the ocean waters. Males collect the eggs laid by the females for their mouth brooding. After incubation, males deposit the fry in rivers where they spend their early stages of life and subsequently, they migrate back to the coastal waters.

Traumatogenic structures: In this species, its pectoral spine is strongly serrated and traumatogenic. Further, it possesses traumatogenic venom glands in its first dorsal spine (https://www.fishkeeper.co.uk/help-and-advice/freshwater/brackish-water-fish/colombian-shark-catfish).

Arius acutirostris (Day, 1877)

Source: Public domain

Common name (s): Arius.

Global distribution: Irrawaddy and Salween River basins of Myanmar and Kraburi River estuary of Thailand.

Ecology: It inhabits estuaries and lower reaches of tidal rivers.

Biology

Description: In this species, its snout is fleshy, elongated, and acute and is extending beyond the mouth. Mouth is distinctly inferior. Upper portion of rayed dorsal and upper two thirds of adipose dorsal are black tinged; and pectoral, pelvic, and anal fins are stained gray. Upper surface of head is granulated or with roughened lines. It attains a maximum length of about 40 cm.

Food and feeding: Not reported.

Reproduction: There is no breeding report.

Traumatogenic structures: It is a traumatogenic species (https://www.fishbase.in/country/CountryChecklist.php?showAll=yes&what=list&trpp=50&c_code=104&cpresence=Reported&sortby=alpha2&vhabitat=dangerous).

Arius africanus (Günther, 1867)

Common name (s): African sea catfish.

Global distribution: Africa: Estuaries of Madagascar and Pangani River, Tanzania.

Ecology: This freshwater, brackish, and benthopelagic species may commonly enter estuaries.

Biology

Description: It has a maximum total length of about 50 cm.

Food and feeding: Its chief food items include small fishes and invertebrates.

Reproduction: There is no breeding report.

Traumatogenic structures: It is a traumatogenic species (https://www.fishbase.se/summary/1284).

Arius arius (Hamilton, 1822)

Source: Public domain

Common name (s): Threadfin sea catfish.

Global distribution: Indo-West Pacific: India, Bangladesh, Myanmar, Singapore, and South China Sea.

Ecology: This demersal, and amphidromous species is commonly found in brackish water lakes, estuaries, and tidal rivers.

Biology

Description: The head shield of this species has a smooth surface anteriorly and granules and rugae are found posteriorly. A large black spot is present in the adipose fin. Lateral line is found bifurcating at the base of the caudal fin. Caudal peduncle is moderately deep. It has a maximum length of about 40 cm.

Food and feeding: It feeds mostly on invertebrates.

Reproduction: Not reported.

Traumatogenic structures: In this traumatogenic fish, the tip of its dorsal spine is filamentous (Chatla and Padmavathi, 2021).

Arius gagora (Hamilton, 1822)

Source: Public domain

Common name (s): Gagora catfish.

Global distribution: Indian Ocean: India and Bangladesh; and South China Sea.

Ecology: This demersal and amphidromous species is commonly seen in estuaries and tidal rivers and estuaries. It is also known to ascend rivers beyond tidal reach.

Biology

Description: Body of this species is elongated with a depressed head. Mouth is narrow and sub-terminal with three pairs of barbels. Maxillary barbels reach pectoral base. Head has median groove which is pointed posteriorly. Coloration of body is dull brown above and dim white on its flanks and belly. All fins are blackish and the blackish spot is seen in its adipose fin. It attains a maximum length of 91 cm.

Food and feeding: It feeds mainly on small fishes.

Reproduction: Its size at sexual maturity is 30 cm total length. Its fecundity has been reported to range from 43000 to 54000 in April to August. Its breeding season which is significantly correlated with rainfall is from July to August and its single peak spawning month is August.

Traumatogenic structures: In this traumatogenic and venomous fish, both the pectoral and dorsal spines are of almost same length. The stings of this species may be very dangerous (http://www.wildsingapore.com/wildfacts/vertebrates/fish/ariidae/ariidae.htm).

Arius jella (Day, 1877) (= *Pseudarius jella*)

Source: Public domain

Common name (s): Blackfin sea catfish.

Global distribution: Indian Ocean: India and Sri Lanka.

Ecology: This marine, brackish, demersal, and amphidromous species is found mostly in coastal marine waters, estuaries, and tidal rivers.

Biology

Description: It attains a maximum length of 30.0 cm. Other details are wanting.

Food and feeding: It feeds mainly on invertebrates. It is an opportunistic feeder consumes mainly fish eggs, shrimp larvae, and fish eggs. It is also known to feed on the waste food from sewage waters.

Reproduction: Males of this species are mouth brooders and they guard both the fertilized eggs and newly hatched young ones.

Traumatogenic structures: This traumatogenic species has venomous spines at its dorsal and pectoral fins (Chatla and Padmavathi, 2021; https://www.talkaboutfish.com/red-fishes-basses-congers-etc/catfish/).

Arius leptonotacanthus Bleeker, 1849

Common name (s): Smoothspined catfish.

Global distribution: Western Pacific; Southeast Asia.

Ecology: This marine, brackish, and demersal species is common in coastal waters and estuaries.

Biology

Description: Its predorsal profile is elevated at its nape. Barbels of this species are flattened. It attains a maximum length of 25 cm.

Food and feeding: It feeds mainly on invertebrates.

Reproduction: There is no breeding report.

Traumatogenic structures: This traumatogenic species has venomous spines at its dorsal and pectoral fins (https://www.talkaboutfish.com/red-fishes-basses-congers-etc/catfish/; http://www.wildsingapore.com/wildfacts/vertebrates/fish/ariidae/ariidae.htm).

Arius maculatus (Thunberg, 1792)

Image is not available.

Common name (s): Spotted catfish.

Global distribution: Indo-West Pacific: Indo-Australian Archipelago (excluding Australia), Myanmar, India, Sri Lanka, Pakistan, and Bangladesh.

Ecology: This marine, freshwater, brackish, demersal, and potamodromous species is common in coastal marine waters, estuaries, and tidal rivers. Occasionally, it forms schools.

Biology

Description: In this species, head shield is somewhat rugose and its eyes are oval in shape. Caudal fin is forked and pelvic fin is almost nearer to the origin of anal fin. Dorsal head and body are thick brown. Lateral part is gray and is gradually becoming whitish with dark spots near the belly. All fins are fairly red with black ends. Both pectoral and pelvic fins are dark colored above. Adipose fin is with black spot. It reaches a maximum length of about 80 cm.

Food and feeding: It consumes mainly small fishes and invertebrates.

Reproduction: In this species, males have been reported to incubate the eggs laid by the females in their buccal cavity. It is worth mentioning here that, males starve during incubation and they may even swallow one or two eggs in order to maintain their basal metabolism. Hatched fry begin feeding on particles inhaled by the female even in the presence of large yolk.

Traumatogenic structures: Its traumatogenic and venomous dorsal and pectoral fins are with strong, serrated spines (https://www.talkaboutfish.com/red-fishes-basses-congers-etc/catfish/; https://www.fishbase.se/summary/Arius-maculatus; Chatla and Padmavathi, 2021).

Arius madagascariensis (Vaillant, 1894)

Common name (s): Madagascar sea catfish.

Global distribution: Africa: western coast of Madagascar; Mozambique and Tanzania.

Ecology: This marine, freshwater, brackish, demersal, and anadromous species is common in rivers and lakes. It is also found in estuaries.

Biology

Description: Body of this species is moderately compressed. There is a single pair of oval shaped vomerine tooth patches. Dorsal sides of body and head are either gray or grayish brown; and ventral surfaces of body and head are somewhat white. Adipose fin is gray; while dorsal-fin rays are gray, rays of all other fins possess scattered gray-colored melanophores. It reaches a maximum length of about 70 cm.

Food and feeding: Not reported.

Reproduction: There is no breeding report.

Traumatogenic structures: It is a traumatogenic species (https://www.fishbase.in/country/CountryChecklist.php?showAll=yes&what=list&trpp=50&c_code=104&cpresence=Reported&sortby=alpha2&vhabitat=dangerous).

Arius oetik (Bleeker, 1846)

Common name (s): Lowly catfish.

Global distribution: Western Pacific.

Ecology: This marine, brackish, and demersal species is common in coastal waters and estuaries.

Biology

Description: Dorsum of this species is yellowish-brown and belly is white. All fins are yellowish in coloration. Its dorsal fins spine tip has a short filament. It attains a maximum standard length of 23 cm.

Food and feeding: It feeds on benthic invertebrates.

Reproduction: There is no breeding report.

Traumatogenic structures: This traumatogenic species has venomous spines at its dorsal and pectoral fins (https://www.talkaboutfish.com/red-fishes-basses-congers-etc/catfish/; http://www.wildsingapore.com/wildfacts/vertebrates/fish/ariidae/ariidae.htm).

Arius subrostratus (Valenciennes, 1840)

Source: Public domain

Common name (s): Shovelnose sea catfish.

Global distribution: Indo-West Pacific: Philippines and Indonesia; and East Pakistan to Thailand.

Ecology: This demersal and non-migratory species is commonly found in estuaries and tidal rivers. It may also be seen burrowing in the mangrove soil.

Biology

Description: It attains a maximum length of about 40 cm.

Food and feeding: It feeds mainly on invertebrates and its diet includes largely diatoms, sea weeds, polychaete worms, and crustaceans; and detritus.

Reproduction: Males of this species are mouth brooders. In India, its spawning season is between the months of January/April and September/October.

Traumatogenic structures: Its dorsal and pectoral fin spines are traumatogenic and they may inflict painful wounds when the fish is provoked (https://www.guppies.za.net/species.aspx?name=shovelnose-sea-catfish-arius-subrostratus&fid=2700; (https://www.fishbase.se/summary/1295).

Arius venosus (Valenciennes, 1840)

Source: Public domain

Common name (s): Veined catfish.

Global distribution: Indo-West Pacific: Southern China to Indonesia (except Philippines and Australia); and Mozambique Channel to Myanmar.

Ecology: This demersal species inhabits commonly in coastal waters and rarely in estuaries. Possibly, it is known to enter freshwater in some countries.

Biology

Description: It attains a maximum length of 30 cm.

Food and feeding: It feeds mainly on invertebrates and small fishes.

Reproduction: Not reported.

Traumatogenic structures: It is a traumatogenic species (http://www.wildsingapore.com/wildfacts/vertebrates/fish/ariidae/ariidae.htm).

Aspistor quadriscutis (Valenciennes, 1840)

Image credit: Jeppe Kolding (Reproduced with permission).

Common name (s): Bressou sea catfish.

Global distribution: South America: Atlantic coastal rivers.

Ecology: This benthopelagic species is commonly seen in the muddy bottoms of rivers, estuaries, and coastal waters.

Biology

Description: The rostral region of this species has three pair of barbels viz. two pairs are on its lower jaw and one pair is on the posterior end of maxilla; and a prominent saddle-shaped bony plate is seen in front of its dorsal fin. Body is gray-yellow or yellow dorsally and it is white ventrally. It has a maximum length of about 50 cm.

Food and feeding: It feeds mainly on bottom-living invertebrates.

Reproduction: Males of this species are mouth brooders as other species of this family.

Traumatogenic structures: It is a traumatogenic species (https://www.fishbase.de/summary/956).

Batrachocephalus mino (Hamilton, 1822)

Source: Public domain

Common name (s): Beardless sea catfish.

Global distribution: Indo-West Pacific: Myanmar, Malaysia, Indonesia, Thailand, Brunei Darussalam, India, Sri Lanka, and Bangladesh.

Ecology: This demersal species is commonly found in coastal waters, estuaries, and tidal rivers.

Biology

Description: In this species, head-shields rough, corrugated, forming a cross over occiput. Its body coloration is silvery or light brown depending upon state of preservation. It is darkest along back and upper lobe of caudal fin. It attains a maximum length of 25 cm.

Food and feeding: It feeds mainly on invertebrates and small fishes.

Reproduction: There is no breeding report.

Traumatogenic structures: It is a traumatogenic species (https://www.fishbase.in/country/CountryChecklist.php?showAll=yes&what=list&trpp=50&c_code=104&cpresence=Reported&sortby=alpha2&vhabitat=dangerous).

Cathorops belizensis (Marceniuk and Betancur, 2008)

Image credit: Alexandre P. Marceniuk (Reproduced with permission).

Common name (s): Belize sea catfish.

Global distribution: Western Central Atlantic: Belize City, Belize.

Ecology: This brackish and demersal species is also found inhabiting mangrove areas.

Biology

Description: This species has a very long and depressed head and its snout is short. Shield on top of head which is long and highly granulated is covered with a thin skin. At the center-top of head, there is a deep, well-defined groove. The bony process located between head shield and dorsal fin base is fairly short and wide behind. Eyes are of moderate size. Two pairs of nostrils are present. Chin and both jaws possess a total of three pairs of barbells. Mouth is narrow and lower jaw is arched; and lips are thick. Females possess larger tooth patches. Gill openings are found restricted to the sides; and gill rakers are 19–20 on first arch and 16–19 on second arch. Tail fin is deeply forked. It attains a maximum length of 25 cm.

Food and feeding: Not reported.

Reproduction: There is no breeding report.

Traumatogenic structures: In this Traumatogenic species, its dorsal fin spine is equal in length to pectoral fin spine; and rear edges of pectoral spine are with short, inconspicuous serrations (https://www.fishbase.se/summary/Cathorops-belizensis; https://www.fishbase.se/summary/Cathorops-belizensis).

Cephalocassis jatia (Hamilton, 1822)

Source: Public domain

Common name (s): River catfish.

Global distribution: Asia: India, Bangladesh, and Myanmar.

Ecology: This marine, freshwater, brackish, benthopelagic, and amphidromous species is found largely in rivers and estuaries.

Biology

Description: In this species, there are three pairs of barbels. Body is bluish-ark dorsally and belly is pale. All fins are yellowish. Lower caudal lobe is gray colored. Adipose fin has a dark spot. It reaches a maximum length of about 30 cm.

Food and feeding: It feeds largely on small fish and benthic invertebrate animals.

Reproduction: Not reported.

Traumatogenic structures: In this traumatogenic species, its pectoral spines are as long as dorsal spine which is very strong (https://www.fishbase.in/country/CountryChecklist.php?showAll=yes&what=list&trpp=50&c_code=104&cpresence=Reported&sortby=alpha2&vhabitat=dangerous).

Cinetodus froggatti (Ramsay and Ogilby, 1886)

Image credit: Dave Wilson Museums Victoria https://creativecommons.org/licenses/by/3.0/au/

Common name (s): Small-mouthed salmon catfish.

Global distribution: Oceania: Roper River system of, Australia; and Fly-Strickland and Digul river systems of New Guinea Australia.

Ecology: This freshwater, brackish, and demersal species inhabits mainly slow-flowing, often turbid rivers, brackish estuaries, and coastal waters.

Biology

Description: It is a large dark bluish-gray to blackish catfish with a pale underside. It attains a maximum length of 48 cm.

Food and feeding: It feeds mainly on gastropod and bivalve mollusks. It swallows the food whole.

Reproduction: There is no breeding report.

Traumatogenic structures: In this traumatogenic species, the spines of its dorsal and pectoral fins are venomous and dangerous (https://www.fishbase.se/country/CountrySpeciesSummary.php?c_code=036&id=10401; https://fishesofaustralia.net.au/home/species/2757).

Cochlefelis burmanicus (Day, 1870)

Source: Public domain

Common name (s): Not known.

Global distribution: Asia: Myanmar and Thailand.

Ecology: This freshwater, brackish, benthopelagic, and amphidromous species is largely found in freshwaters and tidal rivers.

Biology

Description: It attains a maximum length of 40 cm.

Food and feeding: Not reported.

Reproduction: Not reported.

Traumatogenic structures: It is a traumatogenic species (https://www.fishbase.in/country/CountryChecklist.php?showAll=yes&what=list&trpp=50&c_code=104&cpresence=Reported&sortby=alpha2&vhabitat=dangerous).

Galeichthys peruvianus (Lütken, 1874)

Image is not available.

Common name (s): Peruvian sea catfish.

Global distribution: Southeastern Pacific: Peru.

Ecology: it is a marine, demersal species largely seen in coastal waters.

Biology

Description: This species has the following characteristic features: lean body and head; well-tapered posterior body; pointed caudal fin lobes; longer upper lobe of caudal fin; long and slender caudal peduncle; and the presence of a silvery to reddish stripe on lateral body.

Body and head of this species are slender. Body is well-tapered posteriorly. Top of head is smooth, and is covered with thick skin. Snout is rounded and mouth is moderately wide. Jaws are with wide bands of small teeth. Among its three pairs of barbels, top-most is not reaching pectoral fins. Caudal fin lobes are pointed and its upper lobe is longer. Caudal peduncle is long and narrow. Upper body coloration is blue-black to red-brown and belly is white. A silvery white to cream stripe is seen from shoulder to middle of tail base. Fins are dark. It attains a maximum length of 35 cm.

Food and feeding: Not reported.

Reproduction: Not reported.

Traumatogenic structures: It is a traumatogenic species (https://www.fishbase.se/summary/13493).

Hemiarius dioctes (Kailola, 2000)

Image credit: Tim Berra. Museums Victoria, https://creativecommons.org/licenses/by/3.0/au/

Common name (s): Warrior catfish.

Global distribution: Western Central Pacific: New Guinea and Australia.

Ecology: This marine, freshwater, brackish, and demersal species inhabits largely in coastal and lower mangrove areas, estuaries, and lagoons.

Biology

Description: In this species, its snout is prominent; jaw teeth are exposed; and mouth is wide. Head is somewhat venulosus and its shield is almost smooth. Eyes are small. There are no gill rakers on posterior face of first two arches; and gill opening is wide. Body coloration is orange-yellow. Lips, inside mouth, and body are usually covered with orange or yellow mucus; and teeth are usually reddish. It attains a maximum length of 120 cm.

Food and feeding: Not reported.

Reproduction: Not reported.

Traumatogenic structures: It is a Traumatogenic Species (https://fishbase.mnhn.fr/Summary/SpeciesSummary.php?ID=60002&AT=Warrior+catfish).

Hexanematichthys sagor (Hamilton, 1822)

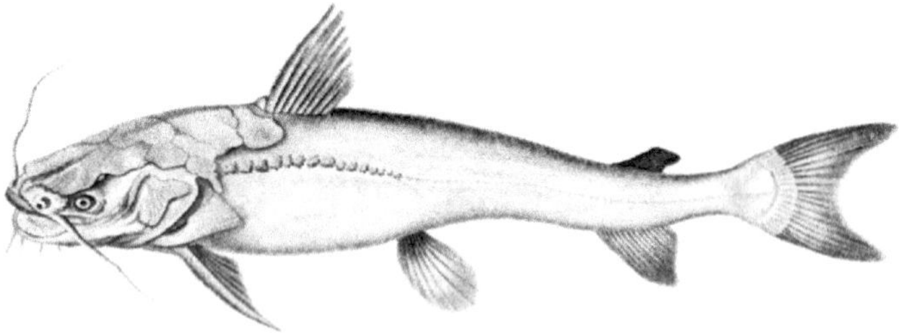

Source: Public domain

Common name (s): Sagor Catfish.

Global distribution: Indo-West Pacific: Pakistan, India, Bangladesh, Myanmar, Thailand, Malaysia, Singapore, China, and Philippines.

Ecology: It inhabits in marine coastal and estuarine areas (with muddy substrate) and also in tidal rivers.

Biology

Description: In this species, head shield is striate and rugose. Supraoccipital spine is broadly rounded and is almost hemispherical. Barbels are flattened and strap-like. Barbels of upper jaw are invariably longer. Body flanks are provided with alternating light and dark vertical bands. It has a maximum length of about 50 cm.

Food and feeding: Feeds on invertebrates and small fishes.

Reproduction: In this species, males and females fish could be distinguished only when they reach sexual maturity. Length at first sexual maturity of the fish is 25 cm. This species has year round spawning. Based on the weight and length, the fecundity of this species may range from 60 to 145 ovaries.

Traumatogenic structures: In this traumatogenic species, venomous spines are seen at the dorsal and pectoral fins (https://www.talkaboutfish.com/red-fishes-basses-congers-etc/catfish/; http://www.wildsingapore.com/wildfacts/vertebrates/fish/ariidae/ariidae.htm;https://www.fishbase.in/country/CountryChecklist.php?showAll=yes&what=list&trpp=50&c_code=104&cpresence=Reported&sortby=alpha2&vhabitat=dangerous).

Nemapteryx armiger (De Vis, 1884)

Image credit: Gerald R. Allen / Western Australian Museum. Museums Victoria https://creativecommons.org/licenses/by/3.0/au/

Common name (s): Threadfin catfish.

Global distribution: Western Pacific: Southern New Guinea and northern Australia.

Ecology: This benthopelagic species dwells in coastal waters and estuaries; rivers within tidal influence; and river deltas.

Biology

Description: In this species, coloration of body is golden brown or grayish pink. Its dorsomedian head groove is elliptical posteriorly and deep. There are long barbels and length of maxillary pair is about 50% of the length of the fish. Bases of chin barbels are staggered and well separated. The first two gill arches lack gill rakers. It attains a maximum length of 40 cm.

Food and feeding: It is an opportunist feeder and consumes mainly crustaceans and small fishes. Aquatic insects, plant materials, and fish scales are consumed occasionally.

Reproduction: Not reported.

Traumatogenic structures: In this traumatogenic species, dorsal fin spine is with a short filament at all ages (https://www.fishbase.in/summary/Nemapteryx-armiger.html).

Nemapteryx caelata (Valenciennes, 1840)

Source: Public domain

Common name (s): Engraved catfish.

Global distribution: Indo-Pacific: Indo-Australian archipelago (except Philippines and Australia), India, Sri Lanka, Pakistan, Bangladesh, Myanmar, and Thailand.

Ecology: This demersal and amphidromous species is living mostly in coastal waters. It is known to enter tidal rivers and brackish waters and is routinely ascends into freshwaters.

Biology

Description: Head profile of this species is fairly concave at nape. There is a head shield which is granulated. Body has a metallic blue coloration. While dorsal fin's tip is blackish, its filament is black in coloration. Pectoral, pelvic, and anal fins are dark; tail fin is pale; and its large, dusky adipose fin may bear a large black blotch on its upper half. It attains a maximum length of about 60 cm.

Food and feeding: It feeds mainly on invertebrates and small fishes.

Reproduction: There is no breeding report.

Traumatogenic structures: In this traumatogenic species, very strong, thick, and coarsely granulated spines are seen in both dorsal and pectoral fins. Tip of its dorsal spine is also produced into a long filament (Chatla and Padmavathi, 2021; https://www.fishbase.in/country/CountryChecklist.php?showAll=yes&what=list&trpp=50&c_code=104&cpresence=Reported&sortby=alpha2&vhabitat=dangerous).

Neoarius leptaspis (Bleeker, 1862) (= *Sciades leptaspis*)

Image credit: Freshwater Fish Group & Fish Health Unit, Murdoch University. Museums Victoria, https://creativecommons.org/licenses/by/3.0/au/

Common name (s): Salmon catfish.

Global distribution: Western Pacific: northern Australia and central-southern New Guinea.

Ecology: This demersal species occurs in muddy coastal waters, estuaries, lakes, and rivers. It is largely seen in upper reaches of streams, brackish estuaries, and lower reaches of rivers.

Biology

Description: It attains a maximum length of 60 cm.

Food and feeding: It feeds largely on crustaceans, mollusks, and fishes. Insects, terrestrial fruits, aquatic plants, and bottom detritus are also taken by these fishes occasionally.

Reproduction: It has been reported to breed in coastal lagoons and flooded swamps during the late dry season and early wet season (i.e., September–January). Males serve as mouth brooders and they incubate the eggs in their mouth until they hatch and protect the emerged fry for about 1 month.

Traumatogenic structures: It is a traumatogenic species (https://www.fishbase.in/summary/Neoarius-leptaspis.html).

Neoarius pectoralis (Kailola, 2000)

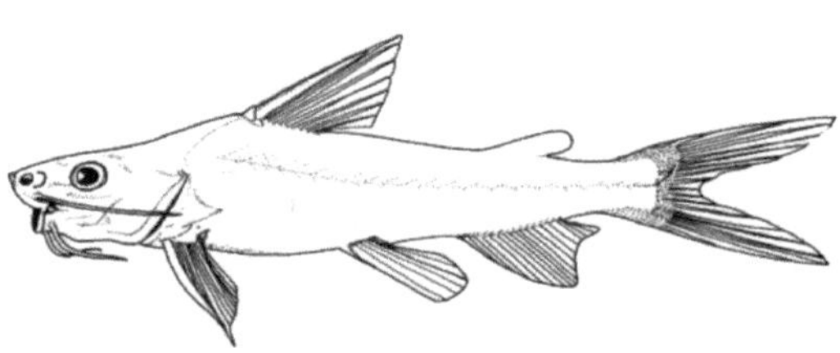

Image credit: Patricia Kailola Museums Victoria, https://creativecommons.org/licenses/by/3.0/au/

Common name (s): Sawspine catfish.

Global distribution: Western Central Pacific: Irian Jaya and Australia. It also found in southern Papua New Guinea.

Ecology: This marine, brackish, and demersal species is largely found in coastal waters, estuaries, and rivers within tidal influence. It also inhabits mangroves.

Biology

Description: In this species, head and eyes are large. Head shield is granular with a long narrow median groove. Snout is acutely pointed; lips are fleshy and crenulated; and lower jaw is squarish. Caudal fin is with slender pointed lobes. Dorsal, pectoral, and inner caudal fin are with distinct black margins. It attains a maximum length of 39 cm.

Food and feeding: Not reported.

Reproduction: Not reported.

Traumatogenic structures: In this traumatogenic species, the inner margin of pectoral-fin spine has large flattened serrations. The venomous spines of dorsal and pectoral fins of this species have been reported to inflict a very painful wound when this fish is provoked (https://www.fishbase.de/summary/Neoarius-pectoralis.html; https://fishesofaustralia.net.au/home/species/2144).

Neoarius utarus (Kailola, 1990)

Common name (s): Northern rivers catfish.

Global distribution: Asia and Oceania: Indonesia and Papua New Guinea.

Ecology: This benthopelagic species is largely seen in main river channels. Lakes and floodplain may also be its habitats.

Biology

Description: It attains a maximum length of 55 cm.

Food and feeding: It feeds mainly on fishes and crustaceans such as *Macrobrachium* and *Caridina.* Aquatic and terrestrial insects, nymphs, and detritus may also serve its food items.

Reproduction: It is a mouth brooder. Ovary or brood mass of this species may contain 15–75 ripe eggs.

Traumatogenic structures: It is a traumatogenic species (https://www.fishbase.in/summary/13127).

Notarius biffi (Betancur and Acero, 2004)

Image is not available.

Common name (s): Chomba sea catfish.

Global distribution: Eastern Central Pacific: Costa Rica and El Salvador; Panama.

Ecology: This benthopelagic species inhabits largely in coastal waters and high salinity estuaries estuaries.

Biology

Description: It has a small mouth and large eyes. Relatively slender caudal peduncle and triangular humeral process are seen. Three pairs of barbels are present. There are no gill rakers on the first two gill arches. It attains a maximum length of 32 cm.

Food and feeding: Not reported.

Reproduction: Not reported.

Traumatogenic structures: It is a traumatogenic species (https://www.fishbase.se/summary/62180).

Notarius bonillai (Miles, 1945)

Image credit: Ray Simpson (Reproduced with permission).

Common name (s): New Granada sea catfish.

Global distribution: Western Atlantic: Colombia.

Ecology: This freshwater, brackish, and demersal species is largely found in turbid water over muddy bottoms in the lower portions of streams, estuaries, and mangrove-lined lagoons.

Biology

Description: Body of this species is elongated and robust with a strongly arched back. Its greatest body depth is at dorsal fin. Snout is broadly rounded. Nostrils are not connected by a furrow. Anterior nostril is fleshy and posterior one is covered by a flap. No fleshy furrow is seen in the middle of head shield. Eye is small and mouth is subterminal and wide with fleshy lips. In this species, there is one maxillary barbel and two mental barbels. Maxillary barbel is not reaching the pectoral fin. Teeth are conical. Gill rakers are present on all arches. At the rear end of the fish, pelvic and anal fins are seen. Adipose fin is present over anal fin base. Caudal fin is strongly forked. Body is naked. Head and dorsum are blackish to blue-black, becoming abruptly whitish on flanks and belly. Fins are uniformly grayish. It has a maximum length of about 80 cm.

Food and feeding: Not reported.

Reproduction: Not reported.

Traumatogenic structures: In this traumatogenic species, its dorsal fin is high with a strong serrated spine. Pectoral fins are about the same size as dorsal fin and are with strong, weakly serrated spines (https://www.fishbase.se/summary/Notarius-bonillai).

Notarius insculptus (Jordan and Gilbert, 1883)

D. R. Robertson; Published under CC0 Public Domain Dedication.

https://treatment.plazi.org/id/03CB8781FFF2FFF82E77FC092A77FACE

Common name (s): Neglected sea catfish.

Global distribution: Eastern Central Pacific: Panama.

Ecology: This marine, demersal species is largely seen in coastal waters.

Biology

Description: The characteristic features of this species include three pairs of barbels; absence of fleshy furrow between the posterior nostrils; absence of fleshy groove in median depression of head; absence of granules or spinulations on the anterior surface of head shield; and absence of gill rakers on rear surfaces of first two gill arches. It attains a maximum length of about 30 cm.

Food and feeding: Not reported.

Reproduction: Not reported.

Traumatogenic structures: It is a traumatogenic species (https://www.fishbase.se/summary/63323).

Notarius neogranatensis (Acero and Betancur, 2002)

Image is not available.

Common name (s): Caribbean sculptured sea catfish.

Global distribution: Western Atlantic: Caribbean coast of Colombia.

Ecology: It is a marine, brackish and demersal species.

Biology

Description: Head of this species is rounded, broad, and slightly flattened. Snout is overhanging mouth. Eye is small. Shield on top of head is very rough and is almost reaching to between eyes. Two pairs of nostrils and three pairs of barbels are present. The barbels are located one chin and one on both jaws. Gill opening is wide. There are no gill rakers on the rear surfaces of the first two gill arches. Tail fin is deeply forked and a complete lateral line is seen. Alcohol preserved fish is brownish above and whitish below. It attains a maximum length of about 40 cm.

Food and feeding: Not reported.

Reproduction: Not reported.

Traumatogenic structures: It is a Traumatogenic species (https://www.fishbase.se/summary/Notarius-kessleri).

Notarius troschelii (Gill, 1863)

Image is not available.

Common name (s): Chili sea catfish.

Global distribution: Eastern Pacific: Mexico to Peru.

Ecology: It is a marine, brackish, and demersal species commonly found in coastal waters. It is possibly restricted to marine environments.

Biology

Description: Body of this species is robust and depressed; snout is prominent; mouth is large; and eyes are medium sized (5.8–8.5 times in head length). Cephalic plate is granulated. Jaw teeth are small, and are set in broad bands. Palate is with four plates of villiform teeth and the lateral ones are larger, triangular, and are extended backwards. Lower branch of the first gill arch has 7–10 gill rakers. Body coloration is dark brown or bluish on back; silvery white on sides and belly; and fins are dusky. It grows to a maximum length of 71 cm.

Food and feeding: Not reported.

Reproduction: Not reported.

Traumatogenic structures: It is a traumatogenic species (https://www.fishbase.se/summary/notarius-troschelii.html).

Osteogeneiosus militaris (Linnaeus, 1758)

Image credit: Staticd, commons.wikimedia. https://creativecommons.org/licenses/by-sa/3.0/

Common name (s): Soldier Catfish.

Global distribution: Indo-West Pacific: Singapore, Malaysia, Indonesia, Malacca, Brunei, Darussalam, India, Bangladesh, Myanmar, and Pakistan.

Ecology: This potamodromous species is found in marine, coastal waters, as well as in estuaries and river mouths.

Biology

Description: Head of this species is strongly depressed and its shield is covered with a soft skin. There are only one pair barbels (maxillary barbels) and these long and stiff barbels are extending to the pectoral fin base. Spines of fins are thin. Body is bluish above; and belly is with white or cream coloration. A dark blue margin is seen on the dorsal and adipose fins. It has a maximum total length of about 40 cm.

Food and feeding: It feeds mainly on invertebrates and small fishes.

Reproduction: Not reported.

Traumatogenic structures: This traumatogenic species has venomous spines at the dorsal and pectoral fins (https://www.talkaboutfish.com/red-fishes-basses-congers-etc/catfish/; https://www.fishbase.in/country/CountryChecklist.php?showAll=yes&what=list&trpp=50&c_code=104&cpresence=Reported&sortby=alpha2&vhabitat=dangerous).

Plicofollis argyropleuron (Valenciennes, 1840)

Image credit: Museums Victoria, https://creativecommons.org/licenses/by/3.0/au/

Common name (s): Longsnouted catfish.

Global distribution: Indo-West Pacific: Northern Australia, southern New Guinea, and South and Southeast Asia.

Ecology: This marine, brackish, and demersal species occurs in inshore waters over soft bottoms in inshore coastal waters and lower estuaries.

Biology

Description: Head of this species is strongly depressed and long and its shield is covered with smooth skin. It has a small mouth and long snout. Tail peduncle is deep. Adipose dorsal fin is dusky in coloration. It has a maximum fork length of 50 cm.

Food and feeding: It feeds mainly on soft-bodied organisms, prawns, detritus, and mud.

Reproduction: Not reported.

Traumatogenic structures: This traumatogenic species has venomous spines at its dorsal and pectoral fins (https://www.talkaboutfish.com/red-fishes-basses-congers-etc/catfish/; https://www.fishbase.se/summary/11739).

Plicofollis dussumieri (Valenciennes, 1840)

Common name (s): Blacktip sea catfish.

Global distribution: Indo-west Pacific: Mozambique and Madagascar to Sri Lanka; India, and Bangladesh; and Myanmar to Sumatra.

Ecology: This marine, freshwater, brackish; and demersal species is commonly found in the coastal waters.

Biology

Description: It has a prominent lateral ethmoid which serves as a shelf between nostrils and eye; two pairs of vomerine tooth plates; and a very short adipose-fin base. Body is dark brown and fine brown pigment specks are seen covering the entire lower surfaces. It attains a maximum length of 85 cm.

Food and feeding: It feeds mainly on invertebrates and small fishes.

Reproduction: Not reported.

Traumatogenic structures: It is a traumatogenic species (https://www.fishbase.se/summary/1287; Chatla and Padmavathi, 2021).

Plicofollis magatensis (Herre, 1926)

No image available.

Common name (s): Luzon sea catfish.

Global distribution: Asia: Philippines.

Ecology: It is a freshwater demersal species.

Biology

Description: In this species, its lateral line is bifurcated at the tail fin base and is almost reaching the lower and upper lobes of tail fin. Base of adipose fin is very short. Origin of adipose fin is above posterior half of anal fin. There is one accessory tooth patch on each side of palate and is positioned posteriorly to vomerine patch. Maxilla is short to moderately long and is shorter than maxillary barbels. Two pairs of mental barbels are seen. The characteristic feature of Plicofollis magatensis is the lateral ethmoid which is not expanded. It attains a maximum length of 45 cm.

Food and feeding: Not reported.

Reproduction: Not reported.

Traumatogenic structures: It is a traumatogenic species (https://www.fishbase.se/country/CountryChecklist.php?resultPage=5&c_code=608&vhabitat=dangerous).

Plicofollis nella (Valenciennes, 1840)

Common name (s): Smooth-headed catfish; shield head catfish.

Global distribution: Western Pacific: South and Southeast Asia, southern New Guinea and northern Australia.

Ecology: It is a marine, brackish, and demersal species.

Biology

Description: Body of this species is elongated and compressed. Head is depressed and head shield is almost smooth and weakly striated. The anterior two gill arches lack inner gill rakers. Body coloration is silvery and brownish dorsally. All fins are blackish or dusky. Adipose fin is with a black margin. It attains a maximum length of 47 cm.

Food and feeding: Not reported.

Reproduction: Not reported.

Traumatogenic structures: It is a traumatogenic species (https://www.fish-base.se/summary/23383).

Plicofollis platystomus (Day, 1877) (= *Arius playstomus*)

Source: Public domain

Common name (s): Flatmouth sea catfish.

Global distribution: Indian Ocean: India and Sri Lanka; Pakistan, Myanmar, and Bangladesh.

Ecology: This marine, brackish, demersal and amphidromous occurs mainly in coastal waters and estuaries.

Biology

Description: It attains a maximum length of 31 cm.

Food and feeding: It feeds mainly on benthic invertebrates.

Reproduction: There is no breeding report.

Traumatogenic structures: It is a traumatogenic species.

(https://www.fishbase.in/summary/1292; https://www.fishbase.in/country/CountryChecklist.php?showAll=yes&what=list&trpp=50&c_code=104&cpresence=Reported&sortby=alpha2&vhabitat=dangerous).

Plicofollis tonggol (Bleeker, 1846)

Source: Public domain

Common name (s): Roughback sea catfish.

Global distribution: Indo-West Pacific: Pakistan, Malaysia, Philippines and Indonesia.

Ecology: This marine, brackish and demersal species is found mainly in coastal waters, and is seldom entering tidal rivers and estuaries.

Biology

Description: It attains a maximum length of 40 cm.

Food and feeding: It feeds mainly on invertebrates.

Reproduction: There is no breeding report.

Traumatogenic structures: The sharp dorsal and pectoral fin spines of this traumatogenic species may inflict painful wounds (https://www.fishbase.se/country/CountryChecklist.php?resultPage=5&c_code=608&vhabitat=dangerous; sharp dorsal and pectoral fin spines can inflict painful wounds).

Sciades sona (Hamilton, 1822) (= *Hemiarius sona*)

Image credit: Robert Mintern–Day, Francis, commons.wikimedia.

Common name (s): Sona sea catfish.

Global distribution: Indo-West Pacific: Pakistan, Thailand, Indonesia, and Polynesia. Also, it is known from the Mekong delta.

Ecology: This marine, brackish, demersal, and amphidromous species inhabits mainly in coastal waters, estuaries, and tidal rivers.

Biology

Description: It attains a maximum length of 92 cm.

Food and feeding: It feeds mainly on invertebrates and small fishes.

Reproduction: No breeding report.

Traumatogenic structures: This traumatogenic species has sharp dorsal and pectoral fin spines which can inflict painful wounds (https://www.fishbase.se/summary/1294; Chatla and Padmavathi, 2021).

Synodontis nigrita (Valenciennes, 1840)

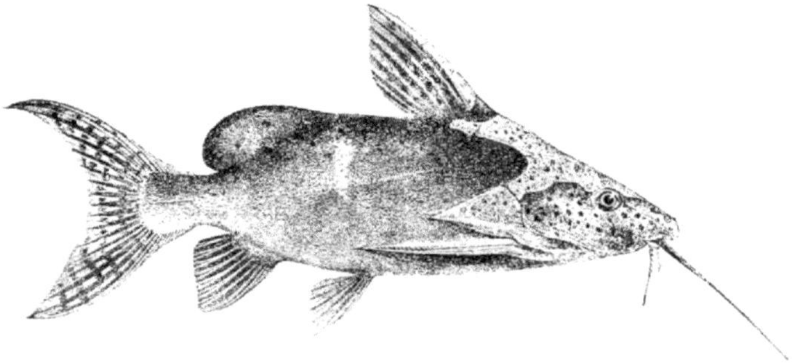

Source: Public domain

Common name (s): False upside-down catfish.

Global distribution: Ghana, Cameroon, Sudan, Egypt, Niger, Chad, Gambia, Nigeria, Mali, and Senegal.

Ecology: It usually dwells in small pools, ponds, and backwaters.

Biology

Description: It attains a maximum length of 17.5 cm.

Food and feeding: This omnivorous species consumes mainly frozen, live, and dried food items. It is also known to relish shelled peas, cucumber, etc.

Reproduction: This species breeds normally in areas with rich microorganisms. It is an egg scatterer and exhibits no parental care.

Traumatogenic structures: This traumatogenic species inflict wounds by its single sharp, stout spine present immediately in front of the soft-rayed portion of the dorsal and pectoral fins (Kapoor and Khanna, 2004).

5.2 CETOPSIDAE (WHALE CATFISHES)

Cetopsis candiru (Spix and Agassiz, 1829)

Image is not available

Common name (s): Whale cat; Candiru-Açu.

Global distribution: South America: Amazon River basin.

Ecology: It is a freshwater, benthopelagic species.

Biology

Description: It has a shallow body especially at the origin of the dorsal fin. Their teeth on the vomer and dentary are of taxonomic importance. They attains a maximum length of 26 cm.

Food and feeding: This species is known for its voracious feeding habit. It has a single row of incisional teeth. It has been reported to feed on dead animals including humans.

Reproduction: There is no breeding report (Lesimann et al., 2020).

Traumatogenic structures: *Cetopsis* sp. is a traumatic fish and it was responsible for the injury of a 2-year-old boy who happened to enter the Madeira River (Lesimann et al., 2020).

Cetopsis coecutiens (Lichtenstein, 1819)

Source: Public domain

Common name (s): Blue whale catfish, blue dolphin catfish, torpedo catfish, and Candiru-Açu.

Global distribution: South America: Orinoco, Amazon, and Tocantins River basins.

Ecology: This benthopelagic species can predominantly swim in large rivers with strong currents.

Biology

Description: It attains a maximum length of 27 cm.

Food and feeding: Its preferred food items include earthworms, prawns, cockles, mussels, etc. It is also believed to be a predator or parasite as it is adapted for feeding both dead animals and larger live fish in the wild. It is also known to feed on mammals.

Reproduction: Unknown.

Traumatogenic structures: Cetopsis sp. has been reported to cause a traumatic injury in a 2-year-old boy who entered Madeira River (Lesimann et al., 2020).

5.3 SILURIDAE–GLASS CATFISHES

Wallago attu (Bloch and Schneider, 1801)

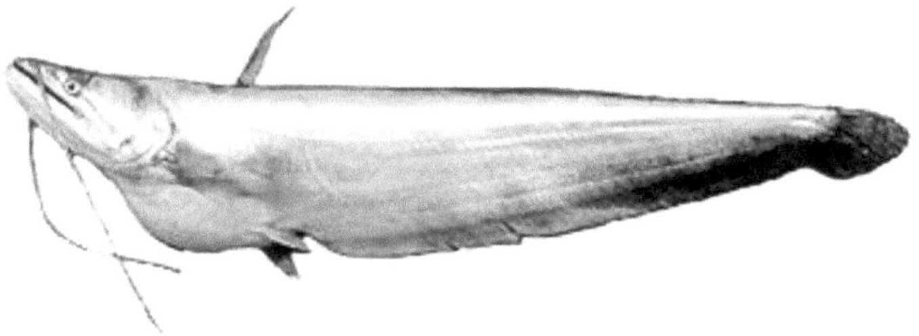

Image credit: Reezwan, commons.wikimedia. https://creativecommons.org/licenses/by-sa/3.0/

Common name (s): Wallago.

Global distribution: Asia: Afghanistan and India; and Pakistan to Viet Nam and Indonesia.

Ecology: It is a demersal and potamodromous species inhabiting tanks, lakes, and large rivers. This predatory catfish is also known to hide under the holes in canals and river banks. It is also known to be associated with slow-flowing water with mud substrate or in deep, still waters. It is also sluggish staying on silty or muddy bottoms in search of food.

Biology

Description: Head of this species is broad, and snout is depressed. Body is elongated and is strongly compressed. Mouth is very deeply cleft. Jaws have teeth which are set in wide bands. Vomerine teeth are present in two small patches. There are two pairs of barbels viz. maxillary and mandibular. Maxillary barbels are found extending to the posterior part of anal fin. There are small eyes. While dorsal fin is small, anal fin is considerably very long. Mandibular barbel is longer than pelvic fin. First gill arch has about 30 gill rakers. It attains a maximum length of 240 cm.

Food and feeding: It feeds largely on crustaceans, mollusks, and small fishes.

Reproduction: It is oviparous species and their pairing is distinct. It is abundant during the warm season and is believed to be a pre-monsoon summer breeder.

Traumatogenic structures: Its large mouth, formidable jaws, and conical teeth are very dangerous and they may help in biting humans when it is handled or provoked. Faint hearted people may be the immediate victims for this venomous and traumatogenic fish (https://www.fishbase.se/summary/10243).

KEYWORDS

- **common name**
- **global distribution**
- **ecology**
- **biology**
- **venomous organs**

CHAPTER 6

VENOMOLOGY OF CATFISHES

ABSTRACT

The various aspects of venomology of catfishes are presented in this chapter. These aspects include venom apparatus of catfishes; envenomation due to catfish stings, and crinotoxins; clinical presentation and manifestations (effects) due to catfish envenomation; case reports relating to envenomation of catfish stings; treatment and management of catfish envenomation; and toxicity and pharmacological properties of catfish venoms.

6.1 VENOM APPARATUS OF CATFISHES

Most of the1000 species of venomous catfish possess a sharp and stout bony spines or stings which are located in front of the soft-rayed portion of dorsal and pectoral fins. These stings are invariably used against their predators for defense. These stings which are the modified fin rays are enclosed by a thin integumentary sheath. The microscopic, venom glands are seen alongside the spines. During the venom delivery, the membrane surrounding the venom gland cells gets torn, thereby releasing the venom into the victim's wound. In other words, the venom from the venom apparatus is injected into the victim. During this period, the stings occur in the hands and feet of the victims like fishermen or beachgoers (Anon., https://www.consultant360.com/article/catfish-spine-envenomation; Anon., https://www.livescience.com/10615-1-000-catfish-species-venomous.html). Toxic secretions of venomous catfishes are of two kinds viz. acanthotoxin and crninotoxin. Acanthotoxin is associated with the venom apparatus present in the dorsal

Biology and Ecology of the Venomous Catfishes. Ramasamy Santhanam, PhD (Author)

and pectoral fins. This venom apparatus is consisting of both spines and venom glands or delivery apparatus. Unlike acanthotoxin, the crinotoxin is produced by the epidermal glands and it gets dispersed over the entire body of the fish (Egge and Simons, 2011). The stings of a common ictalurid catfish, *Noturus gladiator* are given below.

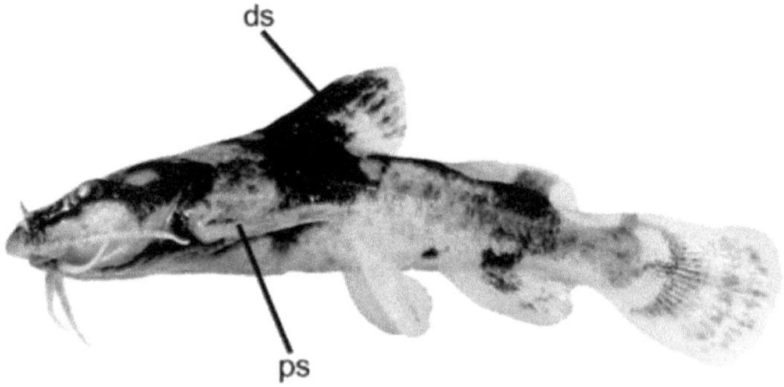

Noturus gladiator showing dorsal and pectoral spine.
Image credit: Jacob J. D. Egge and Andrew M. Simons (Reproduced with permission).

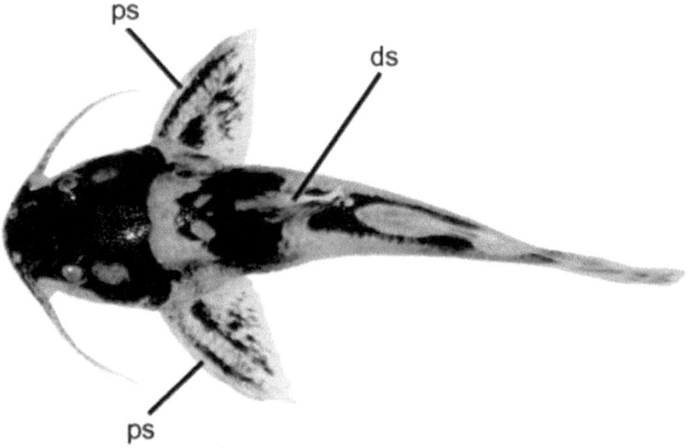

Dorsal view of *Noturus gladiator* showing dorsal spine and pectoral spines.
Image credit: Jacob J. D. Egge and Andrew M. Simons (Reproduced with permission).

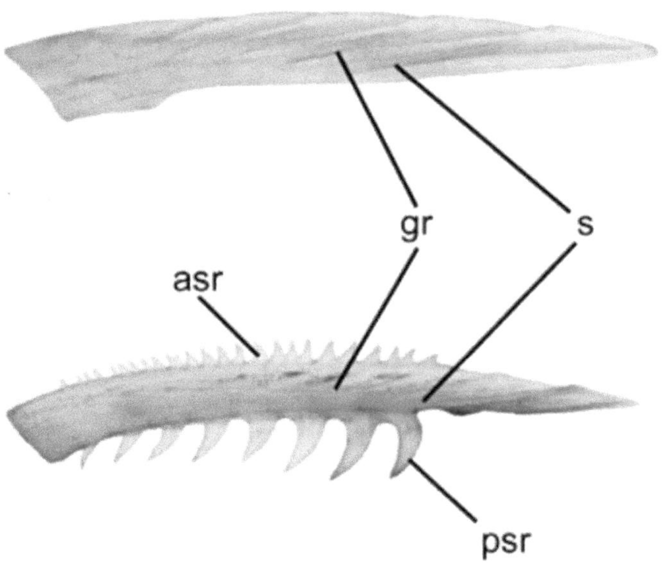

Top: Right pectoral spine of *Noturus lachneri*; Bottom: Right pectoral spine of *Noturus gladiator*; asr, anterior serrations; gr, grooves; s, spine shaft; psr, posterior serrations.
Image credit: Jacob J. D. Egge and Andrew M. Simons (Reproduced with permission).

Venom glands: In some species of catfish, single-celled venom glands may be in larger aggregation, thereby forming full-fledged venom glands which may resemble multicellular glands of terrestrial animals. Such venom glands are normally located around the venomous spines or hard rays of the dorsal or pectoral fins. As there is no common outlet for the above aggregated venom glands they are not termed as multicellular glands (Satora et al., 2008).

Axillary glands: Many catfish species also have secretory glands in the axil of their pectoral fins in addition to their venom glands lining the spinous elements of the fins, such axillary glands which are believed to be a part of the venom apparatus are seen in the catfishes of families like Ictaluridae, Mochokidae, Akysidae, Ariidae, Callichthyidae, and Plotosidae These axillary glands are generally small pouch-like structures which release their venomous secretions through a pore present under the post-cleithral process of the pectoral fin spine base. In certain species, these glands may take a triangular shape. Interiorly, the gland gets divided into many lobes. The secretions of the axillary glands are ichthyotoxic (Wright, 2015).

Venom delivery structures in madtom catfishes: Structures of venom delivery in madtom catfishes: Egge and Simons (2011) reported on the venom delivery structures viz. spine, venom gland, and sting of certain species of freshwater madtom catfishes belonging to the family Ictaluridae which are given below:

Number of stings (serrated spined venom delivery structures) in madtom catfishes

Species	No. of stings
Noturus albater	3
Noturus baileyi	3
Noturus elegans	3
Noturus eleutherus	3
Noturus fasciatus	3
Noturus flavater	3
Noturus flavipinnis	3
Noturus furiosus	3
Noturus gladiator	3
Noturus maydeni	3
Noturus miurus	3
Noturus munitus	3
Noturus placidus	3
Noturus stigmosus	3
Noturus taylori	3
Noturus exilis	2
Noturus hildebrandi	2
Noturus insignis	2
Noturus phaeus	2
Ictalurus punctatus	2
Pylodictis olivaris	2

Source: Adapted from Egge and Simons (2011).

Madtom catfishes with a singer sting (smooth spined venom delivery structure)
Ameiurus natalis
Noturus funebris
Noturus flavus
Noturus gyrinus
Noturus lachneri
Noturus leptacanthus
Noturus nocturnus

Source: Adapted from Egge and Simons (2011).

6.2 ENVENOMATION BY CATFISHES

Envenomation of catfish sting is known to be a frequent cause of morbidity among fishermen, anglers, food processors and aquarists who often handle these fishes. The stings invariably occur in their hands during unhooking of catfish during fishing exercise. Affected victims have reported severe inflammation and pain due to the catfish sting. The wounds caused by such stings may require immediate and judicious care and prophylactic antibiotic administration, and suitable follow-up of the emergency wing (Blomkalns, and Otten, 1999). The venomous catfish have been reported to possess two toxicity mechanisms viz acanthotoxicity and crinotoxicity. Acanthotoxicity is due to the envenomation associated with the spines of pectoral and dorsal fins. During this process, sting penetration and rupture of the venom glandular tissue also occur. On the other hand, crinotoxicity is associated with the secretion of toxins from the entire fish skin (Satora et al., 2008). The number of stings and the amount of venom released by the catfishes determine the severity of the envenomation of the sea fishes.

Envenomation due to catfish stings: In general, venomous catfish have a spine in their dorsal fin and in each of the pectoral fins. Spines of the fins have a series of razor-sharp teeth that act as barbs. These spines normally lie flat against the side of the fish but they get extended when the fish are frightened or agitated (Dart, 2004). The venom glands are seen attached in the leading edges of these spines and are enclosed by integumentary sheaths. During the venom delivery, the venomous spine penetrates into the skin and subdermal tissues of the victims and this activity ruptures the spine integument for the release of the venom of the venom glands into the wound. Catfish envenomations usually occur when the angler of fisherman removes the hook. Though the hands of the fishermen are often inflicted by the catfish sting, anglers are invariably stung on their feet. The venom of catfish is a mixture of dermonecrotic, edema-producing, hemolytic, and vasospastic factors the potency of which is inversely proportional to the size of the fish. The exact venom constituents, however, vary from species to species. Tropical and salt-water species are known to produce the worst symptoms including death of animals and humans (Blomkalns, and Otten, 1999).

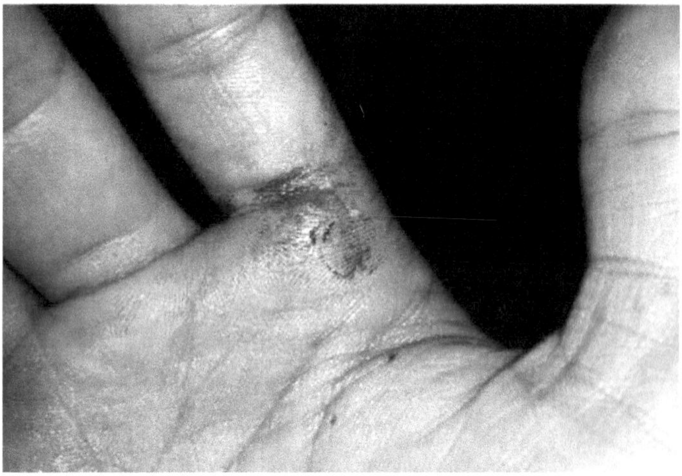

Catfish sting at the base of the index finger.
Image credit: Geoffrey Isbister (Reproduced with permission).

Envenomation due to catfish crinotoxins: The crinotoxicity is believed to be the second toxic mechanism of catfishes. The proteinaceous crinotoxins which are secreted by the tubercle glands of skin of the entire body and fins are believed to be introduced into the wound of the victim during spine envenomation. These toxins have been reported to cause tissue necrosis, muscle fasciculation, and throbbing pain. Crinotoxins are produced by catfishes of the families Ariidae and Plotosidae. Notable species of these families are *Netuma thlassina* and *Plotosus lineatus*, respectively.

Envenomation-associated secondary infection: Freshwater catfishes have been reported to be largely responsible for envenomation-associated secondary infection. This is mainly due to the pollution of their habitats due to the mixing of pathogenic bacteria-rich dirty waters. In freshwater and coastal water habitats of catfish, there are mainly two species of Gram-negative bacteria of the Pseudomonadacea family which are said to be responsible for the secondary infections during fish stings and bites. Aeromonas bacteria may infect wounds inflicted by fish inhabiting both freshwater and salt water. Persons injured by catfish are commonly exposed to bacterial species of *Erysipelothrix, Klebsiella, Escherichia, Edwardisiella, Nocardia, Chromobacterium, Actinomyces, Mycobacterium, Aeromonas*, and *Vibrio.* In humans with reduced immunity, sepsis may result and this is due to infections of *Aeromonas* and *Vibrio* and their toxins in blood and tissues. Another threat due to catfish is erysipeloid which is a skin disease of bacterial infections. *Erysipelothrix rhusiopathiae*, the sources of which are mainly fish and pigs. Erysipeloid occurs usually in fishermen and fish traders and gets a nick name fish-handler's disease. Affected individuals may develop a very strong pain followed by oedema and circumscribed bluish-red skin lesion (Satora et al., 2008).

Clinical presentation and manifestations (effects) due to catfish envenomation

The effects of catfish stings vary from a simple traumatic puncture injury to severe pain associated with envenomation. Localized pain has been reported as the major symptom in most cases but this pain may radiate in severe cases. This pain usually subsides within an hour but it may persist for up to 48 h with more severe envenomation, such as with *Plotosus lineatus*. The local effects displayed by the victims of catfish envenomation include pain, puncture wound, swelling, erythema (redness of the skin), hematoma (collection of blood outside of a blood vessel), paresthesia (abnormal skin sensation including pricking, burning, numbness, etc.), tenosynovitis (inflammation of the fluid-filled sheath that surrounds a tendon), and necrosis (cell injury), etc. On the other hand, the systemic manifestations of catfish envenomation are hypertension (increased heart beats, that is, above 100 per minute), chills, dizziness, vomiting, weakness, agitation, etc. (Blomkalns and Otten, 1999; Satora et al., 2008).

Impact of catfish stings on children and adults: Catfish stings have immediate weakness, pain, and throbbing at the site of injury in the case of both children and adults. Radiation to the proximal extremity is common. In these individuals, the venom is responsible for several inflammatory reaction with erythema, edema, hemorrhage, and tissue necrosis. The severity of these symptoms is highly variable and it largely depends on the species of catfish as well as the amount of envenomation. Symptoms are usually self-limited.

Most serious complications, however, arise from secondary bacterial infection with common organisms including *Vibrio, Aeromonas, Enterobacteriaceae, Pseudomonas, Erysipelothrix, Chromobacterium, Nocardia, Actinomyces, Mycobacterium*, and normal skin flora (Anon., https://www.consultant360.com/article/catfish-spine-envenomation). Catfish venom has also been reported to poison the nerves of a victim and destroys his red blood cells, thereby causing muscle spasms and respiratory distress; severe pain and reduced blood flow (Anon., https://www.livescience.com/10615-1-000-catfish-species-venomous.html). While all freshwater and saltwater catfish species are dangerous saltwater catfish are more harmful than their freshwater counterparts. As far as the size of the catfishes, the smaller ones are often causing more harm to the intruders (Anon., https://www.liveabout.com/finned-by-a-saltwater-catfish-2929252).

Catfish fauna implicated in human envenomation

Venomous spines are believed to be absent in many species of catfish. Although, about 20 catfish families with 1600 species are said to be venomous, envenomations in terms of human injuries have been reported only in few species representing 9–11 families. It is worth mentioning here that about 80% of human injuries are caused by only marine catfish. The families of catfish responsible for human envenomations in different countries are given in the following table.

Catfish families with large scale human envenomations

Ariidae and Plotosidae: Further, the family Ariidae (distributed worldwide) and the fishes of the family Plotosidae of the Indo-Pacific region are very significant as their venom causes local effects such as severe pain, swelling, erythema, pallor, and skin necrosis; and other systemic problems do not appear at the time of injury. However, other complications such as large scale lacerations and abscesses may appear subsequently (Negreiros et al., 2016). Worldwide, primarily fishermen and less commonly water-sports participants are known to be affected by catfish stings. Among the fishermen, saltwater fishermen are often prone to the attack of the potently toxic *Plotosus lineatus* which is known for its frequent envenomations.

This species has been known to cause several health problems including muscular spasm, neurotoxicity, and respiratory distress, and even death in experimental animals (Dart, 2004). Instances of death of fishermen due to myocardial perforation by marine catfish sting have also been common (Haddad Jr et al., 2008).

Important catfish families of aquatic biotopes responsible for human envenomations in different countries

Countries affected	Family
Freshwater	
North America	Ictaluridae
South America	Doradidae and Pimelodidae
Africa, Asia, Europe	Siluridae
Asia	Bagridae and Heteropneustidae
Indo-Pacific	Claridae
Marine	
Worldwide	Aridae
Indo-Pacific	Plotosidae

Source: Adapted from Blomkalns and Otten (1999).

Ictaluridae: Next to the plotosid catfishes, the ictalurid North American catfish are considered to be more venomous. These fishes are said to be largely responsible for secondary bacterial and fungal infections which are introduced through the puncture wound. Further, complications due to these infections and microorganisms have been reported to last for several months (Wright.https://www.sciencedaily.com/releases/2009/12/091210173607.htm).

Siluroidea: Marine catfish injuries often occur as puncture wounds (90%) and lacerations (10%) for which serrated and large stings of catfish fins are mainly responsible. Based on observations, the catfish envenomations and associated injuries may occur in humans in their right hand (37%), right foot (28%), left foot (21.5%), and left hand (21%). This injury pattern is also compatible with the functions of fishermen, walkers, and bathers (Haddad Jr and Martins, 2006). The common species of catfish associated with human envenomation in different countries are listed in the following tables.

Catfish of aquatic biotopes implicated in human injuries in different countries

Countries	Species implicated
Freshwater	
North America	*Ictalurus punctatus, Noturus gyrinus,* and *Noturus* spp.
South America	*Centrochir crocodili, Lithodoras dorsalis,*and *Pterodoras granulosus, Pimelodus blochii, Pimelodus maculatus, Pimelodus pictus, Pimelodella* sp., and *Pseudoplatystoma corruscans*
South America	*Pseudoplatystoma fasciatum* and *Pseudoplatystoma reticulatum*
Africa	*Clarias gariepinus* and *Synodontis njassae*
Asia	*Liobagrus reini, Pseudobagrus, Heteropneustes fossilis,* and *Pangasianodon hypophthalmus*
Australia	*Tandanus bostocki*
Indo- Pacific	*Clarias batrachus*
Marine	
Worldwide	*Ariopsis felis*
Western Atlantic	*Bagre marinus*
Southwest Atlantic	*Genidens genidens*
Indo-West Pacific	*Netuma thalassina*
Indo-Pacific	*Plotosus lineatus*
Australia	*Cnidoglanis macrocephalus*

Source: Adapted from Haddad Jr and Martins, 2006

Marine catfish fauna implicated in venomous injuries to humans in different countries

Family	Species	Distribution	Sources
Ariidae	*Ariopsis felis*	Worldwide	Blomkalns, and Otten (1999)
	Bagre marinus	Western Atlantic	Haddad Jr and Martins (2006)
	Genidens genidens	Southwest Atlantic	Haddad Jr. (2008)
	Netuma thalassina	Indo-West Pacific	Blomkalns and Otten (1999)
Plotosidae	*Cnidoglanis macrocephalus*	Australia	Dart (2004)
	Plotosus lineatus	Indo-Pacific	Blomkalns and Otten (1999)

Envenomation of catfish stings and case reports

i) Marine catfish stings-Western Atlantic Ocean coast

A total of 127 injuries due to marine catfish envenomations have been reported from the Western Atlantic Ocean coast in a period of 8 years. Most of these patients reported intense pain during the acute phase of complications due to bacterial and fungi infections and retention of bony fragments in their wounds. In these people, hot water immersion of the affected parts yielded good results in about 20% of the cases. However, secondary infections by foreign bodies could not be prevented by this technique (Haddad Jr. and Martins, 2006).

ii) Ariidae and Plotosidae

Negreiros et al. (2016) reported that the catfish envenomations which occurred in 127 people of S.E. Brazil lead to inflammation and intense pain in them. Though hot water immersion treatment showed some improvement, five patients reported pain and local inflammation which persisted for about a day. The patients were, therefore, admitted in the hospital after about 5 days of the envenomation.

iii) *Plotosus lineatus*

A 52-year-old Chinese-American man whose skin was penetrated by a catfish barb was hospitalized after 10 days. He had erythema and swelling in his right arm. An abscess of his right thumb underwent incision and drainage which was sent already for laboratory analysis for wound culture. Laboratory results showed higher values of white blood counts, and C-reactive protein, the level of which increases normally during inflammation. The patient was given vancomycin and intravenous ampicillin-sulbactam during the period of his stay in the hospital. During this period, his wound symptoms improved considerably wound cultures showed the presence of *Proteus vulgaris* and *Morganella morganii*. After his discharge, the patient was given oral ciprofloxacin and amoxicillin-clavulanate for a period of 10 days. The patient conveyed after a 12-month telephone follow-up about the complete healing of his wound.

Bentura et al. (2017) reported on the envenomation of *Plotosus lineatus* in the fishermen of SE Mediterranean Sea and the observed results are given below.

Envenomation of the catfish *Plotosus lineatus* in SE Mediterranean Sea's fishermen

Type	No. of cases	Percentage
Area of exposure		
Seashore	37	44
Sea	46	55
Injury site		
Foot	18	21
Palm	66	79

Source: Adapted from Bentura et al. (2018).

The envenomation and associated effects of *Plotosus lineatus* recorded with the fishermen of SE Mediterranean Sea are shown in the following table.

Symptoms of Fishermen in SE Mediterranean Sea due to the envenomation of *Plotosus lineatus*

Symptoms	No. of cases
Systemic manifestations	
Hypertension	6
Chills	2
Tachycardia	2
Vomiting	2
Dizziness	1
Local manifestations	
Pain	76
Puncture wound	59
Swelling	28
Erythema	14
Paresthesis	3
Hematoma	3
Necrosis	1
Tenosynovatis	1

Source: Adapted from Bentura et al. (2018).

iv) ***Plotosus canius***

The venom associated with the sting of this fish produced intense pain, swelling, and gangrene of the affected areas of the victims (Auddy, 1996).

v) ***Plotosus* sp.**

The sting of the long-tailed catfish, *Plotosus* sp. leads to a number of symptoms and signs which include intense localized pain, nausea, and the subsequent localized swelling and redness. It is also suggested that soaking of the injured part in hot water yielded relief of pain. However, an important suggestion to the hot water treatment is to immerse more than the affected part. The paresthesia caused by the venom may cause a deficit in interpreting the temperature of the hot water about the injured part. Irrigation of the puncture wound with a bicarbonate solution is also recommended (Pacy, 1966).

vi) ***Tandanus bostocki***

The venom of this species may not only cause extreme pain but also swelling which is mainly due to a hemolytic activity which denotes the damage of red blood cells and release of a fluid into bloodstream in the affected people. The secondary effects due to bacterial and fungal infections are also not ruled out (Anon., https://www.fishbase.in/summary/12099).

vii) ***Pseudoplatystoma reticulatum* and *Pseudoplatystoma corruscans***

The fishermen of the Miranda and Corumbá towns (Upper Paraguay River Basin, S. America) have been reported to be injured by the catfishes of the genus *Pseudoplatystoma*. While 38 individuals of Miranda town reported injuries (30.16%), the affected fishermen were 111 (56.61%). The large and sharp stingers of the aforesaid species carry toxins which have caused pain and necrosis in the affected victims. Further, perforating and/or lacerated wounds, as well as edema, erythema and occasional skin necrosis were also caused. There was also possibility of serious secondary infections caused by pathogenic bacteria belonging to the species of *Vibrio, Pseudomonas Aeromonas,* and *Edwardsiella*. Further, the most common incident was the occurrence of lacerated lesions which were found associated with erythema, edema, necrosis, paresthesia, etc. On the other hand, the patients reported fever, cardiac arrhythmias, or cold sweating very rarely. These symptoms may, however, be associated with subsequent secondary infections or envenomations due to the toxins of the stingers of this catfish (Aquino et al., 2016).

viii) *Pseudoplatystoma corruscans* and *Pseudoplatystoma reticulatum*

Region	Edema	Erythema	Pruritus	L. wound*	P. wound@	Necrosis
Miranda	82%	84%	50%	74%	26%	26%
Corumbá	87%	98%	1%	75%	25%	26%

**Lacerating wound; @ perforating wound.*

Systemic symptoms due to the stings of *Pseudoplatystoma* sp. in the fishermen of Miranda and Corumba regions

Systemic effects	Miranda	Corumbá
Dispnea	2	8
Sneezing	2	0
Cold sweat	22	72
Malaise	2	20
Paresthesia's	16	84
Secondary infection	6	6
Vomiting	2	2
Fever	30	39
Tachycardia	13	54

Source: Adapted from de Aquino1 et al. (2016).

ix) *Pseudoplatystoma corruscans*

Injuries caused by trauma-inducing spotted catfish (*Pseudoplatystoma corruscans*) in the Pantanal of Midwestern Brazil

The fishing community of Pantanal, Midwestern Brazil was affected by many catfish species including the common species viz. *Pseudoplatystoma corruscans*. Injuries in these fishermen were largely due to the mechanical trauma from stingers or teeth, or other traumatogenic structures. They reported that the envenomation due to *Pseudoplatystoma corruscans* was painful as it was responsible for the extreme malaise in the victims in the first few hours, and for the subsequent secondary bacterial and fungal infections. Among the total victims (47) interviewed, 14 fishermen had lesions and 9 fishermen had ulcers (Haddad, 2017).

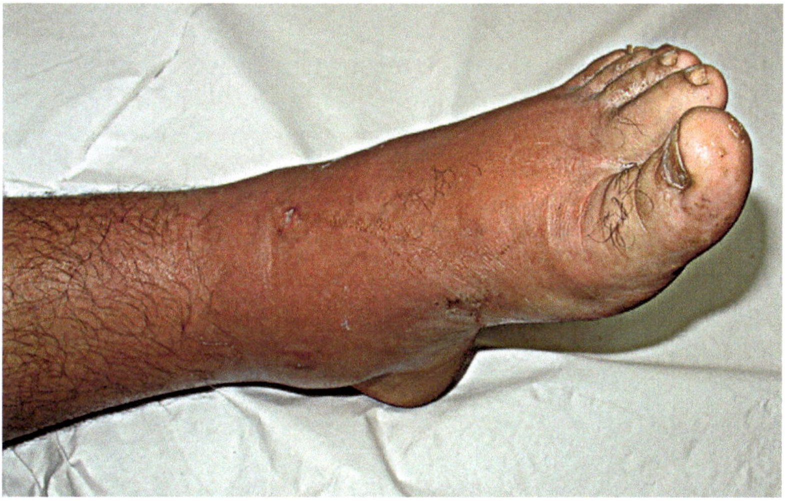

Inflammation and associated pain after a catfish (*Pseudoplatystoma corruscans*) sting in a fisherman of Pantanal.
Image credit: Vidal Haddad Junior.

x) *Synodontis njassae*

While netting the stunned individuals of this species, one individual was nicked on the thumb by the tip of a spine. Within a few seconds, the skin around the nick of the victim swelled into a tight, painful, almost hemispherical lump. In a couple of days, this lump, however, disappeared leaving a small scar at the site of the puncture over 30 years later (Anon., https://www.scotcat.com/articles/article119.html).

xi) *Noturus gyrinus*

A 14-year-old boy was injured by this species in Presque Isle State Park (Erie County, Pennsylvania, USA). He was wounded in the ventral area of his right thumb by one of the pectoral spines. He immediately experienced a burning "hot pin" pain at the puncture site. Bleeding was, however, minimal with two small drops. Burning pain quickly extended to his entire thumb which started to swell and become reddened. By about 30 min, the swelling and redness increased considerably while pain remained constant (McKinstry, 1993).

xii) *Ictalurus punctatus*

A 40-year-old recreational fisherman visited to the emergency clinic about 3 h after he sustained an injury in his left hand and this was due to a freshwater channel catfish. On examination, the volar aspect of his left hand was found imbedded with a spine. He had severe pain, numbness, and dizziness. Subsequent observation showed the presence of a 4 × 3-mm foreign body which was imbedded in the soft tissues midway between the fourth and fifth metacarpal bones. The spine was then removed and the patient was advocated a 7-day course of cephalexin 500 mg four times a day. The patient reported after about 6-months that his hand had healed completely without further complications (Blomkalns and Otten, 1999).

xiii) *Pimelodus maculatus* and *Pimelodella* sp.

In all rivers and lakes of Brazil, these tramumatogenic species have been found to be responsible for all injuries (de Aquino et al., 2016).

xiv) *Pimelodus pictus*

A 26-year old person was admitted in the emergency ward of the hospital due to acute pain and swelling in his light longer pain. It was reported that about 12 h before he had sustained a penetrating injury to his dorsal digit while he was handling the catfish *Pimelodus pictus*. Subsequently, he reported worsening pain, erythema, and swelling at the affected digit. On admission, he had a temperature of 40°C. This patient was given broad -spectrum antibiotic therapy. After an initial stage of improvement, he showed recurrent fever and worsening erythema. Subsequently, a timely debridement was made for his suspected necrotizing fasciitis. After further examination, debridement was limited to the removal of devitalized tissue and copious irrigation of the wound bed. The patient underwent daily debridement for about 48 h with which he defervesced and his white blood cell count became normal. The incision was closed progressively and he was subsequently discharged after the hospital day of 11. With the assistance of aggressive operational therapy, he regained his hand functions (Carty et al., https://journals.lww.com/plasreconsurg/Citation/2010/11000/Digital_Catfish_Envenomation_Mimicking_Necrotizing.28.aspx).

xv) *Pseudoplatystoma corruscans* and *Pseudoplatystoma reticulatum*

It is worth mentioning here that 24% of fishermen of Pantanal, South America had anterior injuries by the stingers of the catfish of the genus *Pseudoplatystoma* which are known for their large, serrated spines on their pectoral and dorsal fins. The stings of these fishes were associated with lacerated or perforating wounds, edema, erythema and skin necrosis. The wounds caused by these species may also lead to serious infections due to pathogenic staphylococci and streptococci, such as *Vibrio, Aeromonas, Pseudomonas* and *Edwardsiella* genera. The injuries caused by these species formed an important public health problem in the continental Brazil (de Aquino et al., 2016).

xvi) *Clarias gariepinus*

A total of seven cases of envenomation with this African catfish venom were reported in Poland. All these people were suspected to be in contact with the stings of the pectoral fins of this species. The affected people were staff of supermarkets, and the injuries occurred during the fish preparation for sale. In all these cases, the injuries were found to be puncture wounds in the right hand in the case of five victims and left hand in two cases. Though intense pain was the main symptom in all the cases, other inflammation manifestations, such as erythema and odema erythema were also found to be common in these patients (Satora et al., 2008).

xvii) *Heteropneustes fossilis*

In Asia, fishermen are afraid of this fish. If they happen to find this species in their nets, they discard this fish immediately with great care. Several cases of human envenomation have been recorded for this species, and the poison of this species poison is believed to be fatal. When this Asian stinging catfish is provoked, the spines of their dorsal and pectoral fins become extended and deliver the venom into the victim who is in contact with this fish. Several cases of human envenomation have been recorded for this species. The foot of victims has been reported to be the most commonly stung part of the body followed by hand injuries by this species (Anon., U.S. Fish and Wildlife Service, 2017; Thalgaspitiya, et al., 2017).

Case report 1: It has been reported that the tissue extracts of this stinging catfish induced favorable cardiotonic activity in isolated hearts of frogs. On the other hand, envenomation of this species is to be considered seriously. The epidermal toxic secretions of this species are known for their rich protein and lipids and poor carbohydrate content. In all admitted cases, the injuries due to this species were mainly of puncture only. The wounds were noted on their left hand in two victims and on right hand in eight people. The injured people experienced intense pain, local oedema, erythema, dizziness, and numbness of the site. In addition, other symptoms such as arterial hypotension, tachycardia, respiratory distress, and unusual sensations like tingling and pricking were also recorded in five cases. These envenomations were considered as moderately serious accidents (Satora et al., 2008).

Case report 2: A 33-years-old aquarist was admitted in the hospital for his painful wound which was caused by this Stinging catfish. The fish stung him in the hand while cleaning the aquarium (Anand et al., 2005).

Case report 3: A 17-year-old aquarist visited the District Hospital in Limanowa (Southern Poland) with an injury in the fifth finger of his right hand and this according to him was due to stinging catfish of his aquarium tank. He was stung by the spine of the first ray of the pectoral fin of the catfish which was about 15 cm in length. The patient also complained of intense pain at the site of the injury; and numbness and dizziness. Further, on examination, a puncture wound of about 1 mm in diameter. was observed at the base of his finger. He had 37.2°C; and heart rate and blood pressure were 96 and 110/60, respectively, at the time of his admission. However, ESR, complete blood count, electrolytes, enzymes, and coagulation profile were all found to be normal (Satora et al., 2005).

xviii) *Pangasianodon hypophthalmus* (= *Pangasius sutchi*)

A 51-year-old Malay man visited the emergency wing of a hospital due to severe pain in his left hand and this according to him was due to a puncture wound caused by the spine of a catfish. The spine which was that of a pectoral fin got penetrated into the dorsal side of his left hand. Though the spine was removed at the spot with the help of his friend, he still complained restricted hand motion, swelling, and heavy pain. His vital signs were, however, normal. Dorsal aspect of his left hand was markedly swollen and the swelling was pulsatile. A spine of the pectoral fin of the stinging fish was seen embedded very strongly in his upper left hand. Blood was found oozing slowly from the wound and his fingers were not cyanotic. His finger movements and sensation were normal. He did not show much difficulty in flexing his wrist joint (Ahmad et al., 2013).

xix) *Bagre marinus*

The dorsal and pectoral fins are equipped with long venomous spines and the stings can cause both severe trauma (lacerations and punctures) and envenomation in humans (Haddad Jr and Martins, 2006). The stings of this species can cause both severe trauma (lacerations and punctures) and envenomation in humans (Haddad Jr and Martins, 2006). Its envenomation manifests locally as oedema, erythema, intense pain, and paleness; and cutaneous necrosis which occurred occasionally. Complications may, however, occur as secondary bacterial and fungal infections when the broken stings form foreign bodies in the wounds. The bacterial infections associated with the envenomation of this species are usually due to the species of *Staphylococcus* and *Streptococcus* and rarely by bacteria of the species of *Vibrio, Aeromonas, Pseudomonas* and *Edwardsiella.* Fungal infections are mainly caused by the species, *Sporothrix schenkii* (Haddad Jr and Martins, 2006).

xx) *Genidens genidens*

The *in situ* external necroscopic examination of the affected person revealed a small perforating wound to the upper left chest of the victim. The autopsy showed a perforation in the victim's left ventricle and there was an intense intrathoracic hemorrhage. Further examination of the patient showed no water in his lungs. The myocardial perforation in the victim's left ventricle could have been caused by the putative spine(s) of this species (Haddad, Jr et al., 2008).

xxi) *Plotosus lineatus* and *Ariopsis felis*

Two patients affected by the stingers of these species showed laceration, bleeding, oedema, and erythema in the area of bites. No systemic symptoms and signs had, however, developed. Neurovascular examinations of the hands were normal (Dorooshi, 2012).

xxii) *Noturus* spp. (Madtoms)

Glands present at the base of the extremely sharp dorsal and pectoral fins of these species secrete venom which is incorporated in the slime and cells that make up each spine. Contact of the spine of this species has been reported to burn like fire (Anon., https://www.gameandfishmag.com/editorial/identifying-fish-that-sting-and-maybe-kill-you/339189).

xxiii) *Cetopsis candiru* or *Cetopsis coecutiens*

A 2-year-old boy who entered the Madeira River with his mother to bath on the Madeira Riverbank got traumatogenic injury by the bite of *Cetopsis* sp. which according to Lesimann et al. (2020) was suspected to be either *Cetopsis candiru* or *Cetopsis coecutiens*. As the child began to cry and shout, the mother safely brought her son out of the river and saw a candiru fish clinging on the child's leg. When the candiru left the child, it left a deep circular lesion of 3×3 cm^2 size on the left calf of the affected child. The mother applied saline solution for asepsis (prevention of development of bacteria, viruses, and other microorganisms) and she used to clean the wound 3 times a day.

xxiv) Case reports of unidentified species

a) Case report: Murphey et al. (1992) reported on two cases with serious secondary infections associated with catfish spine-related injuries. For these infections, microbes such as *Pseudomonas* spp, *Vibrio* spp., and *Aeromonas hydrophila* were found to be largely responsible.

b) Case report: Chan et al. (2010) stated on the catfish envenomations to 54 patients whose hands were affected by the stings of catfish. All these patients experienced heavy pain at the site of the sting. Other manifestations observed include numbness (n = 13), wound swelling (n = 28), and erythema (n = 13). However, there were five complicated cases viz. one case of abscess formation in the sting area, three cases of cellulitis, and one case of subacute tenosynovitis. All these five patients were examined for the treatment with parenteral antibiotics (Anon., https://www.consultant360.com/article/catfish-spine-envenomation).

c) Case report: A 39-year-old suffered a puncture wound to the dorsum of his right hand while he was removing a catfish from the fishing hook. This injury leads to an immediate severe pain and swelling, which spread almost to his whole limb. Erythema was noted at the site of the puncture. He was initially treated with a 10-day course of cephalexin by his physician. His pain, edema, and erythema decreased subsequently. Initially, no attempt was made to extract the foreign body. Further, no radiograph was taken on initial presentation. In the subsequent months, the patient, however, had recurrent pain and swelling and till 6 months he developed a knot on the dorsum of his hand. This swelling got ruptured spontaneously and blood-stained fluid was found discharged.

On further examination, the patient was with a painful swollen right hand and wrist, with erythema around the suture line. A plain radiograph revealed the presence of a residual foreign body between the heads of the third and fourth metacarpals. Magnetic resonance imaging revealed gross inflammation of the tenosynovium of the wrist https://www.wemjournal.org/article/S1080-6032(03)70540-7/fulltext-fig2. The patient was, therefore, admitted, and treatment was begun with intravenous antibiotics (levofloxacin, amoxicillin/clavulanate). He was taken to the operating room, where the previous incision was opened followed by wound lavage. A white pasty material and inflammation were encountered on opening of the dorsum of the hand between the third and fourth metacarpals. Extensive inflammation prevented the isolation of the foreign body. Needle localization was unsuccessful. As surgical manipulation may lead to dangerously close to the volar neurovascular bundles, the search for the spine was abandoned, and the wound was left open. The wrist capsule was subsequently opened through a separate dorsal incision and this lead to the appearance of a white pasty, granular material involving the edematous tenosynovium and capsule. Histological examination showed chronic inflammation with granulation tissue and a marked histiocytic response. The tissue which was sent for microbiologic examination did not grow any organisms during routine cultures. Further, specialized cultures for vibrio and fungi were negative. The patient, however, responded well to the surgery and antibiotics, and he was discharged on a 6-month course of oral rifampin and clarithromycin.

In the follow-up visit 2 months after the debridement, the patient reported that he still suffered from occasional bouts of pain. In spite of continued antibiotic therapy, he was not symptom free. On further examination, his wrist was found swollen to a lesser degree. Another attempt was, therefore, made to isolate the foreign body to remove the only obvious nidus of infection. His hand was re-examined 15 months after the initial injury and 2 months after the first surgical procedure. A pasty white substance which was similar to what was seen in the wrist during the last procedure was again encountered. The catfish fin which was embedded in the dorsal third web space was removed. The patient was symptom free for the next few months, during which time his antibiotic therapy was also discontinued. However, the patient returned after 5 months with a complaint of recurrent pain. Though the symptoms improved with the use of antibiotics, they continued to recur on cessation. The infectious disease consultant advocated a prolonged course of antibiotics, targeting the mycobacteria, particularly Mycobacterium marinum, using a combination of rifampin and clarithromycin. Though the patient returned to work he still suffered from pain. His wrist remained intermittently swollen, and he remained on antibiotics (Ajmal et al., 2003).

d) Case report: Severe tissue necrosis is invariably causes due to the epidermal secretion viz. skin toxin and the venoms of spines of the dorsal and pectoral fins of catfish. The marine fishermen hands are the common sites of catfish stings. Mann and Werntz (1991) stated about two cases affected by catfish stings in their hands. In one cases, amputation was done for the gangrene of the long and small fingers. Symptoms in these cases were all due to the dermonecrotic, edema-promoting, hemolytic, vasospastic, and lethal components of the venoms and skin toxins of venomous catfish species. Local anesthesia was given for these patients to get rid of vasospasm and pain. Subsequently, intravenous antibiotics were prescribed. Wounds associated with swelling, pain, erythema, and cyanosis were well irrigated to clean out residual toxin and certain cases, and debridement was also administered for the removal of retained spine fragments or necrotic tissue.

e) Catfish spine envenomation in a 2-year-old boy: A commercial fisherman, was trying to free a catfish from his crab trap. When the fish fell, its dorsal spine punctured his 2 -year-old child's ankle leaving a wound. After several attempts, the father cut the spine 1 cm above the skin with a pair of pliers. After the 1 h of this incident, the boy was sent to the emergency wing of the hospital and he had developed pain, erythema, and edema. The child was sedated and the spine was removed intact. The wound was irrigated copiously with saline and it was left to heal. The patient was discharged with oral clindamycin. The child's father informed that the wound of his child healed without any complications (Grams and Mularoni (Anon., https://www.consultant360.com/article/catfish-spine-envenomation).

f) Case report: Reichert et al. (2017) reported on a catfish injury with a 7-year-old boy who was with penetrating trauma. His injured forearm was found to retain foreign body. The initial imaging conveyed transection of the radial artery, but under a controlled setting, this foreign body was found to compress the artery, however, without any vascular injury. No vascular repair was needed after the removal of foreign body.

g) Microbiology case study with a 51-year-old male with swelling in his finger: A 51-year-old man sustained an injury to his left third finger from a catfish sting while he was fishing in a brackish water area of Florida. As his finger developed swelling, he was hospitalized. As the swelling got infected, he was prescribed a couple different courses of antibiotics for about a week. As his finger became more swollen even after this treatment, he was admitted for 3 days on vancomycin/cipro. His MRI report showed prominent tenosynovitis of the left 3rd finger tendon sheath, albeit his second, fourth, and fifth digits were also involved. Though surgical debridement was recommended, he preferred to pursue medical management. He was started on cipro/doxy/clarithro. Unfortunately, after 3 weeks, his left third finger was "burst open" with a draining sinus tract and purulent discharge. He was examined by infectious disease and orthopedic surgery and subsequently, he consented to radical tenosynovectomy for putative mycobacterial chronic tenosynovitis (Rogers and Wojewoda, https://labmedicineblog.com/2018/05/15/microbiology-case-study-a-51-year-old-male-with-swelling-in-his-finger/).

h) Case report: A 14-year-old had been noodling (hand fishing) for catfish with his bare hands and a catfish caused abrasions to the dorsal surface of his right hand before escaping. He also sustained an injury and a deep laceration in his right forearm due to a submerged rusty barrel in which the catfish was placed. He had difficulty in moving his right hand and wrist due to this incident. A thorough examination of his injured right upper extremity showed superficial abrasions at the dorsum of his hand and fingers. A deep laceration of 12 cm long could also be noticed on the extensor surface of the mid right forearm. There was, however, no active bleeding, and his hand was perfused distally. But he was unable to extend his fingers or wrist. The forearm, wrist, and hand radiographs showed soft tissue injury only, without any foreign bodies (Courtney et al., 2013).

i) Case report relating to secondary infections: The catfish spine envenomations may often lead to most serious long-term complications involving bacterial infections. One series of case reports, they reported serious infections required ray amputations in two patients; and another necessitated digit amputation as some patients reported wound progression to dry gangrene. Pathogenic microorganisms included in these case reports are the species of *Aeromonas, Vibrio, Nocardia, Sporothrix, Klebsiella, Erysipelothrix, Chromobacterium, Actinomyces, Edwardsiella,* and *Mycobacterium.* It was observed that patients reported in these case reports were more susceptible to infections from *Vibrio*. It was also reported that harmful species of *Vibrio* and *Aeromonas* were mainly responsible for saltwater infections and freshwater infections, respectively (Blomkalns, and Otten, 1999).

j) Case reports: A series of 83 catfish sting cases have been reported from the inland hospitals (Belzoni and Indianola) and a coastal hospital (Pascagoula). The type and severity of injury, as well as the type of treatment and its effectiveness, have been described for all these cases. Oral cephalosporin has been reported to be adequate in treating early infections in all these cases (Das et all., 1995).

k) Case report: A case of a radial artery laceration injury which is only secondary to any catfish sting was presented. According to this case report, a cartilaginous spine was removed vessel wall was repaired. But after a period of 6 months, this patient was found to show minimal symptoms like cold intolerance (Ferlic et al., 2003).

l) Case report: Three patients who were affected by embedded freshwater catfish spines were examined. In each case, the spine was identified radiographically and was then removed. All these patients completed a course of cephalosporins and recovered without infection (Zeman, 1989).

m) Case report: This report deals with the death of a 39-year-old male fisherman of São Sebastião town, São Paulo State, Brazil. He had a myocardial puncture due to a catfish sting and this incident happened while he was holding a catfish-filled net against his thorax. When he was trying to remove the net from the water, he was agitated in the water for a minute or two, and then submerged under the water's surface. He was immediately rescued from the water by his companions, but he expired within minutes. He had suffered a catfish sting to the left anterior hemithorax. The *in situ* external necroscopic examination of the corpse revealed a small perforating wound to the upper left chest of the victim which suggested an injury caused by a piercing and cutting instrument. The autopsy revealed a perforation in the victim's left ventricle and intense intrathoracic hemorrhage (hemothorax) (Haddad et al., 2008).

n) Case report: A healthy 5-year-old white girl had a catfish spine infection in lake water. The dorsum of her left foot was hit by a sharp spine of a catfish. When she was taken to the emergency wing of a hospital, her 2-cm open wound was sutured first. Radiograph examination showed the presence of a foreign body which was nothing but a 3-cm-long catfish spine and the same was removed. She was subsequently treated with intravenous clindamycin and cefazolin. She had fever up to 103°F, and the affected foot had swelling and erythema. As she was not alright with fever of 105°F and she had tachycardia too. The wound also showed serosanguinous discharge. The wound discharge when cultured showed the presence of microbiota such as *Enterobacter cloacae, Morganella morganii, Aeromonas hydrophila, and Edwardsiella tarda.* Blood cultures were, however, found to be negative. The antibiotic regimen needed to be changed to intravenous ceftriaxone, clindamycin, and gentamicin. The patient was then without fever and with considerable reduction in the swelling and redness of the affected portion. She was advised intravenous antibiotics for an additional period of 2 weeks at her home (Eiland and Salazar, 2006).

6.3 TREATMENT AND MANAGEMENT OF CATFISH ENVENOMATION

Risk factors associated with catfish sting: The predisposing factors relating to the catfish stings include improper handling of live or dead catfish at the fishing site; and swimmers, surfers, divers, and beachgoers who have no fair knowledge about the habitats of catfish.

First-aid measures: The first-aid measures for any catfish stinger include removal of all foreign material; irrigating the affected site subsequently, with a clean liquid; warm water immersion of the injured part; and other pain management measures.

Symptoms of catfish sting: The severity of symptoms of catfish sting may largely depend on factors such as species of catfish, amount of toxin injected, and the impact of catfish toxin on the victim. The symptoms of catfish stings may be of different kinds such as cuts, injuries, cuts, and lacerations which happen normally on the arms and legs of fishermen while they are involved in fishing or angling. Other associated signs include severe bleeding from punctured wounds; rashes on the skin (usually red-colored welts) and bruising; swelling of the wounds; sweating, nausea, and headache; and shortness of breath.

Diagnosis of catfish sting: A catfish sting is usually diagnosed by the physical examination of the affected region of the body by the experienced physicians; careful observation of the symptoms exhibited by the affected individual; and complete blood tests including blood count (CBC) and electrolytes.

Catfish sting-associated complications: The complicated symptoms which may arise due to a catfish sting may include the development of secondary bacterial or fungal infections associated with the bleeding and skin ulceration; allergic reaction of anaphylaxis type; acute respiratory problems; and rare fatalities.

Treatment methods for catfish sting: The treatment methods for catfish sting may include:

i) The first line of treatment relates to the careful removal of stingers/ spines with the help of a pair of tweezers
ii) As the use of very hot water (up to 50°C) is known to reduce the potency of the venom, the wound is soaked for about an hour in this hot water. This method may also help to reduce the intensity of pain and vasoconstriction (ice is not recommended as it is reported to increase pain)
iii) After the removal of spines, the affected region is cleaned with water and soap and the wound is left uncovered to heal
iv) Topical antiseptics may be used to help prevent wound infection
v) Pain relieving medications (like acetaminophen) may be used if required
vi) Tetanus vaccination may be done on the advice of a physician.

Medical emergency: In case of medical emergencies, the following treatment measures may be adopted

i) Cardiopulmonary Resuscitation (CPR) may be provided as breathing assistance if needed
ii) Blood pressure needs to be stabilized
iii) Pain control medications like steroids may be given
iv) Proper follow-up care and check-ups may be attempted in order to monitor for any signs of further infection
v) Local emergency helpline may be contacted for any unwanted situation without any delay.

Prevention of catfish sting: In order to reduce the incidence of catfish stings, the following may be considered

i) Awareness may be created among fishermen and fisherwomen on the safe handling of ctifh in general and smaller-sized fishes in particular
ii) Beach-based warnings from health officials and lifeguards or may be strictly adhered
iii) Wearing protective clothing (wet skin or diver's suit) to swim in catfish infested waters

iv) Protective footwear may be worn while walking on beach
v) Stingers from the body may be removed using hand gloves
vi) Marine life explorers may be advised to keep first-aid kit with them whenever they undertake underwater investigations
vii) Unfortunately, no antivenom is presently available for injuries relating to catfish stinging.

Source: Anon., https://www.dovemed.com/diseases-conditions/catfish-sting/#

Treatment and management for fish stings

Catfish stings are believed to be painful for few hours and this pain can be controlled to a greater extent with the immersion of the injured part in hot water for about 90 min. Further, this method may also help in the disappearance of inflammation during the first 24 h. As the catfish venom is not stable at high temperatures and it degrades in the heat, the pain may reappear once the hot water immersion is stopped. Further, it has been reported that hot water immersion may promote immediate vasodilatation in the wound site (Haddad Jr and Martins, 2006). Since the retained spines in the wound may be radiolucent; debridement and irrigation of the wound are also suggested. Satora et al. (2008) reported on the use of parenteral analgesics which is said to control pain. In this connection, tetanus prophylaxis may be given. It is also to be noted that antibiotic therapy depends largely on the immune status and age and of the patient; the time interval between injury and application of drug; and the presence and type of foreign bodies.

Treatment measures for stingers of individual catfish species

Plotosus lineatus: Most injuries associated with this species occur in fishermen who are trying to pull the fish off lines. Affected person needs to first thrust his hand in to fairly very hot water where its potent toxin is said to be heated and diluted. If things get worse may be rushed to a local hospital with an aid who is working with this fish (Anon., https://www.scotcat.com/factsheets/plotosus_lineatus.htm; Anon., https://www.nps.org.au/australian-prescriber/articles/managing-injuries-by-venomous-sea-creatures-in-australia).

Bentura et al. (2018) reported on the treatments given to victims due to this species in SE Mediterranean Sea.

Method of treatment	No. of victims
Oral analgesics	51
Antibiotics	3
Anti-tetanus prophylaxis	46
Hot water immersion	36
Orthopedic consultation	3
Wound disinfection	84

Source: Adapted from Bentura et al. (2018)

Clarias gariepinus* and *Heteropneustes fossilis: Treatment and management of envenomation due to these species include:

i) First-aid treatment: Immersion of the injured part in about 50°C water for a minimum period of 30 min. This exercise may not only relieve muscle spasms and intense pain, it may also inactivate the venom if and when present in the wound.
ii) Application of radiographs using X-rays or gamma rays at the inspection for foreign bodies.
iii) Use of parenteral analgesics to control pain.
iv) Prophylactic short course of oral antibiotic therapy with trimethoprim–sulfamethoxazole.
v) Tetanus prophylaxis for needy patients and observation of residual deficits in motor or sensory functions, if any.
vi) Antibiotic management: This management practice is depending on factors such as the immune status and age of the patient; the interval between the injury and the presentation; and the presence of a foreign body. It is always better if the patients are admitted for observation in order to avoid complications.
vii) In the case of injury inflicted by these catfish, infection is known to develop in a period of 3 months in the affected persons. Hence, the patients are advised to have check-ups.

Source: Satora et al. (2008)

Clarias gariepinus: Treatment for accidents relating to this species is consisting of wound debridement, and subsequent removal of the broken fin ray remains; immersion of affected parts in hot water for about 40 min;

and applications of antibiotic therapy and tetanus prophylaxis. Satora et al. (2008) reported that such treated wound healed slowly in a period of 1–2 months. By these treatments, no residual deficits in sensory or motor functions were observed. Aid was normally given by first-contact physicians and the Toxicological Centre was subsequently consulted.

Heteropneustes fossilis

Case report 1: The first-aid treatment for the stings of this species was the removal of all foreign material and irrigating the site with any clean liquid followed by warm water immersion and pain management. For the inspection for foreign bodies in the wound, Radiographs were employed. The victims were provided with a trimethoprim–sulfamethoxazole based oral antibiotic therapy for a brief period as a prophylactic measure. Tetanus prophylaxis was given when indicated. If no residual deficits in motor or sensory functions after the patients were observed for 24 h, they were discharged. Over 2 months, the wounds were found healed slowly by second intention (Satora et al., 2008).

Case report 2: The victim was provided tetanus anatoxin and antibiotics after irrigating and debriding of the wound. In order to relieve the pain, an opioid analgesia and local anesthesia were also given to the affected person. The wound of the hand was found healed and the patient had no significant complaints after a period of 2 months (Anand et al., 2005).

Case report 3: The wound region was infiltrated with 1% lidocaine for the removal of foreign bodies. Subsequently, the wound was also well irrigated and cleansed with antiseptic solutions and it was left open. A plain radiographic study made in of the injured area removed all retained barbs and other foreign bodies. Then the affected portion of the hand was immersed in hot water of 45°C for about 50 min. For inactivating the venom in the wound, the victim had a prophylactic short course of oral antibiotic therapy with Trimethoprim–sulfamethoxazole. After 24 h of continuous observation, the victim was discharged from the hospital. As injuries caused by the catfish may lead to delayed appearance of infection, the patient was advised for be admission for observation at the time of need. Over the next month, the wound was found healed slowly by the second intention. In a period of 2 months, the wound was found to be completely healed without any further complication. The motor and sensory function of the hand was also normal (Satora et al., 2005).

***Plotosus lineatus* and *Ariopsis felis*:** The affected limbs of the victims were immersed in hot water (temperature below 50°C) which led to a gradual reduction and elimination of pain in about 30 min. In order to avoid secondary infection, intravenous and intramuscular injections with the antibiotics were administered. For this purpose, cefazolin 1 g (intravenous, every 6 h) and tetanus toxoid 250 IU (intramuscular) were used. Plain radiographs were also taken to check the possible presence of remnants of fish bite in limb and this was found to be normal. The wounds were subsequently irrigated with warm sterile water and 1% povidoneiodine in solution. Wounds were kept open in order to heal from secondary infection. The patients were observed for incidence of systemic symptoms for 24 h and were then discharged from hospital with order of oral antibiotics. At the follow-up, the wounds were found healed slowly by second intention and they had regained the baseline level of hand functions. No residual deficits in motor or sensory functions were also observed (Dorooshi, 2012).

Unidentified catfish: Affected victims were provided irrigation, exploration, and culture of the wounds as well as immunization against tetanus. Immunocompromised individuals and victims with chronic illness and hepatic disease were at unusually high risk of infection due to pathogenic Vibrio and Aeromonas species and such patients were treated with antibiotics. In contrast, victims with late wound care, punctures, fever, or signs of sepsis showed high risk for secondary infection and they were provided with antibiotics. One of the following combinations is advised for infections of different levels: (i) a tetracycline, a-lactamase-stable penicillin and an aminoglycoside; or (ii) a tetracycline and a broad-spectrum, a-lactamase-stable lactam antibiotic (Murphey et al. (1992).

6.4 CHARACTERISTICS OF CATFISH VENOMS

The venom of catfish contains both multiple protein fractions and non-proteinaceous component. Further, its potency is largely inversely proportional to the catfish size and it is largely related to odema-producing and vasospastic hemolytic, and dermonecrotic factors. However, less is known about the detailed composition and potency of catfish venoms (Egge and Simons, 2011; Satora et al., 2008).

Chemistry of catfish venoms: The molecular weight of the toxic peptides of catfish venoms generally fall within the size range of 1020 kDa although some putative toxins of 4050 kDa have also been reported in some Arius spp. On the other hand, toxin with the highest molecular weight, 110 kDa has been recorded in the venom electrophoretic profile of certain siluriform species. It is worthy of mention here that the rapid loss coloration of a natural predator when the latter is injected with catfish venom extracts with toxic peptides of such high molecular weights (Wright, 2015). The differences in venom function of catfish compared to the venomous terrestrial organisms are largely due to the reduced diversity of toxic peptides of their venoms. Detailed examinations of the proteins in catfish have been limited to the venoms of a toxic marine species (*Plotosus canius*,) and a freshwater species (*Ameiurus catus*). The hemolytic and neurotoxic properties of *Plotosus canius* venom were mainly to its 15 kDa protein toxin which is also said to be responsible for causing mortality. On the other hand, the venom of *Ameiurus catus*, has two to eight toxic proteins of 10 kDa molecular weight (Wright, 2012).

Venom characteristics of catfish species

Ameiurus catus: Venom of this species has been reported to possess two to eight fractions which showed mortality in experimental mice. In the injected largemouth bass, *Micropterus salmoides* which is a carnivorous freshwater gamefish, the two putative toxic peptides of fin spine extracts produced a significant toxic effect (Wright, 2015).

Ameiurus nebulosus: Aqueous spine extract of this species has been reported to produce contraction in the small intestine of the guinea-pig. Similar effects were also recorded with the mouse gastrointestine (Barthó et al., 2018).

Arius maculatus: The biochemical properties of the toxic skin mucus of this species have been studied. The protein contents of the soluble and insoluble fractions were 9.34 and 12.64 μg/g, respectively; and the values of total lipid and carbohydrates were 0.005 and 0.08 μg/g, respectively. Of the total 17 amino acids of the skin mucus, leucine and aspartic acid showed high values (9.8 and 9.0 mole %, respectively) and cysteine was the lowest amino acid both in the soluble and insoluble fractions with 0.01 and 0.02 mole %, respectively. The soluble and insoluble fractions of the skin mucus showed hemolytic activity with the values of 25 and 50 μg/ml, respectively; and antibacterial activity against pathogenic *Escherichia coli* and *Pseudomonas aeruginosa* (Manivasagan et al., 2009).

Arius thalassinus: This species has been reported to produce toxin from the skin and venom glands of the pectoral fins. Its skin toxin displayed cholinergic vasoconstrictor activity in umbilical arteries. Further, the cholinergic vasoconstriction was also confirmed in sheep renal arteries. Skin toxin treatment of arterial specimens has also been reported to prostaglandins such as 6-keto-PGF1 alpha, PGE2, and TXB2 (Al-Hassan et al., 1986).

Cathorops spixii: The sting venom of this species showed the presence of 13 proteins which have been reported to cause more harmful effects, such as changes in arteriolar wall diameter, venular stasis, and hemorrhage. The unique protein 65-kDa Wap family protein (WAP65) isolated from this species showed inflammatory activity (Ramos et al., 2012).

Clarias batrachus and Heteropneustes fossilis: The tissue extracts of these species have shown cardiotonic activity in the experimental heart of frog. The extracts from *Heteropneustes fossilis,* however, had a greater influence than those of *Clarias batrachus*. Further, the epidermal secretions of these species were rich in lipids and proteins while carbohydrates were very poor. Among the lipids, the phospholipids were found to be the principal constituents (Satora et al., 2008). The values of the proximate composition of the epidermal secretions and poisonous spine of *Heteropneustes fossilis* and *Clarias batrachus* are shown below.

Proximate composition of toxic epidermal secretions and poisonous spine extracts of *Heteropneustes fossilis* (Hb) and *Clarias batrachus* (Cb)

Item	Epidermal secretions		Spine extracts	
	H.f	*C.b*	*H.f*	*C.b*
Proteins	0.7	1.4	0.4	0.2
Carbohydrates	0.04	0.07	0.1	0.1
Lipids	54.5	8.3	19.1	9.5
Water content	92.1	82.1	47.8	46.9
Dry weight	7.4	18.2	52.5	53.1

(Values in mg/100 mg of fresh weight; lipids as mg/100 mg of dry weight).

Source: Adapted Venkaiah and Lakshmipathi (2000).

Hypostomus plecostomus: The toxins PC, I and II isolated from the spine extracts of this species showed lethal and cytotoxic activities (Pandey and Upadhyay, 2020; Sarmiento et al., 2015).

Ictalurus catus: The fin venom of this species has been reported to contain at least 2–8 mouse lethal fractions and two dermonecrotic fractions. The molecular weight of the lethal fractions was about 10,000 and the pI of two fractions was 3:8 and 7:8. The lethal activities of this venom were found to be stable at 4°C (Burnett, 1975).

Pimelodus maculatus: The crude venom of the sting of this species was found to contain about 100 g of protein. At doses from 30 to 100 g/animal, this venom induced notable edematogenic and nociceptive effects apart from vascular permeability alterations. Furthermore, this crude venom has also been reported to reduce the contraction force *in situ* frog heart, and induce important changes in the levels of the enzyme Creatine kinase (CK) and its isoenzyme viz. Creatine kinase–myocardial brand (CK-MB) in mice (Sarmiento et al., 2015).

Plotosus canius: The mucus and spine extracts of this species showed proteins with molecular weight in the range of 7–21 KDa suggesting that the venom of this species boasts a assorted mixture of peptides, protein, and pharmacologically active components (Prithiviraj and Annadurai, 2010). In the experimental animals, the venom of this species produced hypertension and respiratory failure, cardiac arrest, neuromuscular blockage, and the LD50 value was found to be 3.9 mg/kg (ip) in male albino mice. Further, non-competitive neuromuscular blocking activity of cardiac tissues was also shown by the Toxin-PC of this species (Ziegman and Alewood, 2015).

Plotosus lineatus: The amino acid profile of the spine venom extracts of this species showed that serine was dominant (1.98 g/100g) and leucine was the least (0.14 g/100g). The values of other amino acids derived from this species are given below.

Amino acid profile of spine venom of *Plotosus lineatus*

Amino acid	g/100g
Serine	1.98
Aspargine	1.41
Cysteine	1.31
Glycine	1.31
Aspartic acid	1.11
Metheonine	1.03

(Table Continued)

Amino acid	g/100g
Threonine	1.00
Glutamic acid	0.98
Glutamine	0.98
Arginine	0.89
Alanine	0.79
Iso-leucine	0.56
Lysine	0.35
Histidine	0.34
Proline	0.34
Valine	0.34
Tryptophan	0.22
Phenylalalnine	0.20
Tyrosine	0.20
Leucine	0.14

Source: Adapted from Pachaiyappan et al., (2015).

The crude venom of the skin secretion of this species displayed hemolysin and lethal (edema-forming) factors. The molecular weights of these components were found to be 180,000 and 12,000, respectively. Amazing secretory cells, which are more or less similar to the venom glandular cells of the fin stings have also been observed in the epidermis of this catfish (Shiomi et al., 1986).

6.5 TOXICITY AND PHARMACOLOGICAL PROPERTIES OF CATFISH VENOMS

Nowadays, fish venoms are considered as a potential source of physiological tools and pharmacological agents. However, due to the technical difficulties in isolating and storing the venom extracts, these venoms form an untapped source of therapeutic compounds. It is also worth mentioning here that the cardiovascular system is the most affected one due to these fish venoms. Further, most of these venoms have also been reported to possess cytolytic, neuromuscular, and wound healing activities (Anon., https://www.sciencedaily.com/releases/2009/12/091210173607.htm).

6.5.1 TOXICITY OF VENOMS OF CATFISH TO ANIMALS

Venom toxicity: The toxicity of the venom extracts from the pectoral and dorsal stings of ictalurid catfishes was studied by injecting the same into the experimental fish species, *Gambusia affinis*. Extracts of *Noturus exili*s and *Ictalurus melas* were found to be lethal to the fish and those from *Galeichthys felis, Ictalurus natalis, Noturus gyrinus, Noturus nocturnus, Noturus, flavus,* and *Noturus placidus* were found to be quite virulent. The extracts of *Ictalurus punctatus, ictalurus nebulosus,* and *Noturus albater* were innocuous and the venom of *Noturus leptcanthus* and *Pylodictis olivaris* was not at all toxic (Birkhead, 1972). Similar studies have also shown that the crude extracts of the stings of *Ictalurus melas* and *Schilbeodes exilis* were found to be more toxic than that of *Schilbeodes nocturnus* or *Ictalurus natalis*. Further, the extracts of stings from *Schilbeodes leptacanthus* and *Ictalurus punctatus* showed either little or nil effect in the experimental fish (Birkhead, 1967). Wright (2012) reported on the toxicity index of the venoms of catfish species and it was found that the venoms of *Noturus stigmosus* and *Noturus gyrinus* were more toxic with the toxicity index values of 5.06 and 4.56, respectively. However, the venoms of *Ameiurus* sp. were found to be less toxic with a poor range of values (0.25–1.88). The values of toxicity index recorded for other species of catfish are given below:

Toxicity indices in species of catfish

Species	Toxicity index
Noturus stigmosus	5.1
Noturus gyrinus	4.6
Noturus furiosus	4.1
Noturus miurus	3.8
Noturus insignis	2.5
Noturus hildebrandi	2.4
Noturus funebris	2.3
Noturus flavus	2.1
Noturus natalis	1.9
Noturus exilis	1.8
Noturus elegans	1.7
Noturus fasciatus	1.6

(Table Continued)

Species	Toxicity index
Noturus nebulosus	1.6
Noturus catus	1.6
Noturus phaeus	1.4
Noturus albater	1.3
Noturus platycephalus	1.1
Ictalurus punctatus	1.1
Ameiurus brunneus	0.4
Pylodictis olivaris	0.3
Noturus melas	0.3
Noturus leptacanthus	1.1

Source: Adapted from Wright (2012).

Toxicity of catfish fauna to animals: The venoms of certain species of catfish have been reported to display toxicity (cytotoxicity, hemotoxicity, and cardiotoxicity) in experimental animals as shown below:

Toxicities of catfish to animals

Catfish family	Spp. of catfish	Toxicity type
Pimelodidae	*Pimelodus maculatus*	Cardiotoxic
Plotosidae	*Plotosus canius*	Cardiotoxic
Siluridae	*Ompok bimaculatus*	Hemotoxic
Clariidae	*Clarias batrachus*	Hemotoxic
Clariidae	*Clarias gariepinus*	Hemotoxic
Clariidae	*Ictalurus furcatus*	Hemotoxic
Heteropneustidae	*Heteropneustes fossilis*	Hemotoxic
Pangasiidae	*Pangasianodon hypophthalmus*	Hemotoxic
Bagridae	*Mystus bleekeri*	Cytotoxic
Bagridae	*Mystus vittatus*	Cytotoxic
Pangasiidae	*Pangasius pangasius*	Neurotoxic and hemotoxic

Source: Adapted from Pandey and Upadhyay (2020); Sarmiento et al. (2015); Auddy et al. (1995).

Activities of catfish venoms on the freshwater gamefish, Largemouth bass *(Micropterus salmoides)*: The physiological effects of venoms of certain catfish species have been studied on the experimental fish Largemouth

Bass (*Micropterus salmoides*) that was injected with 2 μL/g body weight of crude venom extract. Among the different extracts, the extract of the marine species *Plotosus lineatus* only showed mortality.

Effect of catfish venom on the fish, Largemouth bass

Species	Effects					
	LE	CL	MC	TS	HG	MY
Arius jordani	+	+		+		
Corydoras paleatus		+				
Horabagrus brachysoma	+	+		+	+	
Microglanis iheringi		+			+	
Noturus gyrinus	+	+	+		+	
Pangasius hypophthalmus	+		+			
Pimelodus pictus		+			+	
Synodontis multipunctata		+	+		+	
Plotosus lineatus				+		+

LE, Loss of equilibrium; CL, color loss; MC, Myoclonus; TS, Tetanus; MY, Mortality.

Source: Adapted from Wright (2009).

6.5.2 PHARMACOLOGICAL PROPERTIES OF CATFISH VENOMS

Cardiovascular properties: The venoms of *Heteropneustes fossilis* and *Plotosus canius* have been reported to cause the release of prostaglandins which are responsible for the production of contractile responses in smooth muscle. On the other hand, the contraction of smooth muscle by the venom of *Arius thallasinus* is said to be produced by its activities on muscarinic acetylcholine receptors. Further, the toxin-PC of *Plotosus canius* developed stopping of heartbeat in guinea pig while the venom of *Heteropneuste fossilis* showed inotropic increases both in toad and guinea pig hearts. Furthermore, the venom of *Plotosus canius* has been shown to produce a hypertensive response, while that of *Heteropneustes fossilis* produced a hypotensive effects (Wright, 2017).

Neuromuscular properties: Several catfish venoms have shown potent neuromuscular properties. The crude extracts of *Plotosus canius* are known to affect the muscle contraction in both chick and rat. Similarly, the venom of *Plotosus lineatus* also developed neurotoxic symptoms when intraperitoneal injection of this fish extract was made mice. However, the venom of *Heteropneustes fossilis* did not display any appreciable neuromuscular effects (Wright, 2017).

Cytolytic properties: Almost the venoms of fishes have been reported to exhibit cytolytic properties. Necrosis (death of cells in organ due to injury) has been reported in the envenomation of the mosquitofish (*Gambusia affinis*) with the venoms of ictalurid catfish venoms. However, such necroses were said to be due to the secondary bacterial infections. Further, the venoms of several catfish species have been reported to produce hemolysis in human (*Arius thallasinus*), rat (*Plotosus canius*, rabbit (*Plotosus lineatus* and *Plotosus canius*), mouse (*Plotosus canius*), cow (*Arius thallasinus* and *Plotosus canius*), and sheep (*Arius thallasinus*) erythrocytes. It is also important to note that the venom of *Plotosus lineatus* was cytotoxic to Ehrlich ascites tumor cells (Wright, 2017).

6.5.3 BIOACTIVITIES OF VENOMS OF CATFISH FAUNA

Arius maculatus: The spine venom of this species has been reported to possess hemolytic activity with the erythrocytes of chicken, sheep, and human blood cells. Further, this venom has also shown proteolytic activity against casein and gelatin (Ziegman and Alewood, 2015).

Arius thalassinus: The spine venom of this species has been reported to possess proteolytic activity against casein ((Ziegman and Alewood, 2015). Al-Hassan et al. (1987) reported that the spine venom and proteinaceous gel-like material from its epidermal cells showed significant pharmacological activities in experimental animals. For example, both the acanthotoxins and crinotoxins of this species possessed hemolytic and hemotoxic activities. Further, its crinotoxins which possess novel protein and lipid constituents showed lethal activity when they were intravenously in experimental rabbits at a dose of about 1 mg protein/kg body weight. These crinotoxins are also believed to play a greater role in wound healing.

Clarias batrachus: The tissue extracts from this species have been reported to possess therapeutic activity by inducing cardiotonic activity in isolated hearts of frogs (Satora et al., 2008).

Clarias batraqchus* and *Saccobranchus fossilis: The venoms of these species have been reported to show profound effects on the hematological parameters of the experimental albino rats (Rattus norvegicus). The lethal dose (LD) of venom of *Saccobranchus fossilis* for the intravenously injected rats was found to be 5 and 20 μg/kg for that of *Clarias batrachus* parameters after the injection of venom on both groups of albino rats. The findings revealed that the treated group of albino rat became anemic as the RBCs of treated decreased due to the effect of the catfish venom. Similar type of result was also observed in the catfish species, Arius maculatus. The venom of *Arius thalassinus* has also been reported to contain a complex mixture of components such as a specific hemolytic factor, esterases, prostaglandins, and platelet activating factors, which also had potent pharmacological effects on test animals. The level of hemoglobin was found to decrease in treated group after it is injected with the venom of this species. In regards other hematological parameters, a decreased level of sodium was found to be highly significant ($P \geq 0.001$) in treated test animals with the venoms of *Clarias batrachus* and *Saccobranchus fossilis*. Similarly, the level of potassium in blood serum was found increased in treated group in comparison to the control group. Further, an increased level of calcium was found slightly significant ($P \geq 0.05$) in the treated group in comparison to the control group (Choudhary and Pandey, 2013). The values of hematological parameters recorded are shown in the following table.

Characteristics of the venoms of *Saccobranchus fossilis* (*Sf*) and *Clarias batrachus* (*Cb*) on Hematological parameters in albino rats

Species	Dosage	Na mg/dl	K mg/dl	Ca mg/dl	Hb gm%	RBC*	WBC*
Cb	20 μg/Kg	124.0	11.2	13.8	12.6	3.3	4.3
Sf	5 μg/Kg	120.0	12.1	14.7	7.2	3.2	5.9

*million/mm^3

Source: Adapted from Choudhary and Pandey (2013).

Heteropneustes fossilis: The tissue extracts from this species have been reported to induce cardiotonic activity in isolated hearts of frogs (Satora et al., 2008).

Netuma thalassina: The proteinaceous material secreted by this species has been reported to show wound healing property. The application of epidermal gel preparations to the wounds of mice, rats, guinea pigs, and humans reduced healing time by up to 60%. Further, it accelerated clotting, vasoconstriction, hemolysis, and the agglutination of red blood cells which may be involved in the healing processes (Al-Hassan et al., 1985). Anon.

(https://www.futurity.org/just-your-garden-variety-poisonous-catfish/) and Wright (2012) reported that the toxins in catfish skin secretions are capable of accelerating wound healing in humans.

Plotosus lineatus (spine venom)

Thrombin-like activity: The spine venom of *Plotosus lineatus* has been reported to possess thrombin-like activity with a value of 15.3 NIH/mg and this activity may help in the coagulation of plasma and fibrinogen (Pachaiyappan et al., 2015).

Antibacterial activity: The spine venom of *Plotosus lineatus* showed antibacterial activity by inhibiting the growth of pathogenic bacterial strains viz. *Escherichia coli, Salmonella.typhi, Vibrio.parahaemolyticus, Vibirio.cholerae, Staphylococcus.aureus,* and *Staphylococcus paratyphi* with values of zone of inhibition 4.0, 5.0, 4.5, 5.5, 3.0, and 7.0 mm, respectively (Pachaiyappan et al., 2015).

Hemolytic activity: The crude spine venom of *Plotosus lineatus* has shown high levels of hemolysis in the erythrocytes of chicken, sheep, goat, and human blood and the values of hemolytic activity in these cells were found to be 10 HU, 7 HU, 4HU, and 7 HU (1 HU represents the lysis of 50% of the red blood cells), respectively (Pachaiyappan et al., 2015).

Cytotoxicity: The crude spine venom of *Plotosus lineatus* has shown cytotoxicity of HEp2 cell line (epithelial cells derived from larynx carcinoma) (Pachaiyappan et al., 2015).

Plotosus canius

Antibacterial activity: The mucus and spine extract from this species showed pharmacological activity. The crude of aqueous, acetone, and chloroform extracts was tested against six species of bacteria viz. *Pseudomonas sp, Streptococcus aureus, Vibrio cholerae, Bacillus sp, E.coli,* and *Lactobacillus brevis* (Prithiviraj and Annadurai, 2010).

Antifungal activity: The mucus and spine extract from this species showed pharmacological activity. The crude of aqueous, acetone, and chloroform extracts were tested against five species of fungai viz. *A.flavus, A.niger, Candida albicans, A.oryzae,* and *A.sojae* (Prithiviraj and Annadurai, 2010).

Antioxidant Assay DPPH (1, 1-diphenyl-2-picrylhydrazyl) assay: The mucus and spine extract from this species showed antioxidant activity. The DPPH reaction with both venom samples showed a significant color change pattern. This result confirmed that the presence of antioxidant molecule in the sample (Prithiviraj and Annadurai, 2010).

Hemolytic Assay: The mucus and spine extract from this species showed pharmacological activity. In case of mucus venom sample, aqueous extract was found to be four and its specific hemolytic activity was estimated to be 2.6 HT/mg of protein. For acetone extract found to be five and its specific hemolytic activity was 2.4 HT/mg of protein, for chloroform extract found to be three and its specific hemolytic activity was 3.5 HT/mg of protein (Prithiviraj and Annadurai, 2010).

Pseudoplatystoma fasciatum: The spine venom extracts of this species have been reported to produce severe inflammation in mice (Ziegman and Alewood, 2015).

Siluroidei catfish (Siluriformes): Pandey and Upadhyay (2020) reported that the Toxin PC, Toxin I, and Toxin II of the spine extracts of the fishes of this group showed a wide array of physiological properties.

KEYWORDS

- **venomology of catfish**
- **envenomation**
- **clinical presentation and manifestations**
- **envenomation case reports**
- **venom toxicity and pharmacology**

REFERENCES

Abirami,P.; Arumugam, M.; Giji, S. ; Nagarajan,S. Bio-prospecting of catfish sting venom *Arius maculatus* available along South East coast of India. *Int. J. Pharm. Pharm.* **2014,** 6, 110–111.

Ahmad, R.; Zaini, I.Z.; Mohamed, N.A.N.; Roslanuddin,M.S. Radial artery injury by a catfish (*Pangasius sutchi*) spine sting.. *Southeast Asian J. Trop. Med. Public Health,* **2013,**44, 703–706.

Ajmal,N.; C Nanney,, L.B.; Wolfort,,S.F. Catfish Spine Envenomation: A Case of Delayed Presentation. *Wilderness Environ. Med.* **2003**, 14, P101–105.

Al-Hassan ,J.M.; Thomson, M.; Criddle,K.R.; Summers,B.; Criddle, R.S. Catfish epidermal secretions in response to threat or injury: a novel defense response. *Mar.Biol.* **1985**,88, 117–123.

Al-Hassan,J.M.; Thomson, M.; Ali ,M.; Criddle,R.S. Toxic and Pharmacologically Active Secretions from the Arabian Gulf Catfish (*Arius thalassinus*, Ruppell). *J. Toxicol. Toxin Rev.* **1987**,6, 1–43.

Al-Hassan, J.M.; Thomson, M.; Ali, M.; Fayad, S.; Elkhawad, A.; Thulesius, O.; Criddle,R.S. Vasoconstrictor components in the Arabian Gulf catfish (Arius thalassinus, ruppell) proteinaceous skin secretion. *Toxicon* **1986,** 24, 1009–1014.

Anand, J.S.; Chodorowski, Z; Waldman,W. Hand wound caused by an active sting with a toxin spine of a catfish (*Heteropneustes fossilis*)--a case report. *Przegl Lek.* **2005,** 62, 526–527.

Armbruster,J.W. Global Catfish Biodiversity. *Am. Fish. Soc. Symp.* **2011, 77,** 15–37.

Aquino, G.N.R.; Souza, C.C.; Haddad, Jr. V.; Sabino, J. Injuries caused by the venomous catfish pintado and cachara (*Pseudoplatystoma* genus) in fishermen of the Pantanal region in Brazil. *An Acad Bras Ciênc.* **2016,** 88, 1531–1537.

Auddy ,B.; Gomes, A. (1996). Indian Catfish (*Plotosus canius*, Hamilton) Venom. In: Singh B.R., Tu A.T. (eds) *Natural Toxins 2. Advances in Experimental Medicine and Biology*, vol 391, Springer, **1996.**

Auddy, B.; Muhuri,D.C.; Alam, M.I.; Gomes,A. A lethal protein toxin (toxin-PC) from the Indian catfish (*Plotosus canius*, Hamilton) venom. *J. Nat. Toxins* **1995,** 3, 363–368.

Barthó, L.; Sándor,Z.; Bencsik,T. Effects of the venom of the brown bullhead catfish (*Ameiurus nebulosus*) on isolated smooth muscles, *Acta Biol. Hung*, **2018,** 69, 135–143 .

Baskaradoss,K.; Kumar,P. Observation of acanthotoxicity effect of freshwater catfish *Mystus gulio* from spine and mucus extract. *Int. j. fish. aquat. sci.* **2018** ,6, 31–37.

Bentura, Y.; Altunina, S.; Levdova , I.; Golanic ,D.; Spanierd , E.; Edelistd,D.'Luriea, Y. (The clinical effects of the venomous Lessepsian migrant fish *Plotosus lineatus* (Thunberg, 1787) in the Southeastern Mediterranean Sea. *Clin. Toxicol*. **2018,** 56, 327–331.

Birkhead ,W.S. The comparative toxicity of stings of the Ictalurid catfish genera I*ctalurus* and *Schilbeodes*. *Comp. Biochem. Physiol*. **1967,** 22, 101–111.

Birkhead, W.S. Toxicity of Stings of Ariid and Ictalurid Catfishes. *Copeia* **1972**, 1972, 790–807.

Blomkalns, A.L.; Otten, E.J. Catfish spine envenomation: a case report and literature review . *Wilderness Environ Med* , **1999**, 10, 242–246.
Burnett ,J.W. Catfish [*Ictalurus catus*] fin venom. *Toxicon* **1975,** 13: 399–403.
Carty,M.; Kutz,R.H.; Finley,R.L.; upton,J.; Rogers,G.F. Digital Catfish Envenomation Mimicking Necrotizing Fasciitis. https://journals.lww.com/plasreconsurg/Citation/2010/11000/Digital_Catfish_Envenomation_Mimicking_Necrotizing.28.aspx
Carvalho, T.I.; Klaczko , J.; Slobodian,V. Pectoral-fin glands and delivery apparatus in the catfish genus *Brachyrhamdia* Myers, 1927 (Siluriformes: Heptapteridae). *Pap. Avulsos Zool.* **2021** , 61, https://doi.org/10.11606/1807-0205/2021.61.74
Chan , H.Y.; Chan , Y.C.; Tse , M.L.; Lau, F.L. Venomous fish sting cases reported to Hong Kong Poison Information Centre: a three-year retrospective study on epidemiology and management. *Hong Kong J. Emerg. Med.* **2010,** 17, 40–44.
Chatla,D.; Padmavathi , P. Fish Diversity of Coastal Andhra Pradesh, Southeast Coast of India. *Adv. Anim. Vet. Sci* . **202**1,9,1424–36.
Choudhary, A.; Pandey,P. Effect of catfish venom on haematological parameters in albino rat. *Indian j. fundam. appl. life sci.* **2013**, 3, 9–15.
Clark,E.; Nelson, D.R.; Stoll,M.J.; Kobayashi,Y. Swarming, diel movements, feeding and cleaning behavior of juvenile venomous eeltail catfishes, *Plotosus lineatus* and *P. japonicus* (Siluriformes: Plotosidae). *aqua, Intl. J.Ichthyol.* **2011,** 17, 211–239.
Dart, R.C. *Medical Toxicology;* Lippincott Williams & Wilkins, 2004.
Das, S.K.; Johnson, M.B.; Cohly, H.H. Catfish stings in Mississippi. *South Med. J.* **1995**, 88, 809.
Donascimiento, C.; Lundberg, J.G. *Myoglanis aspredinoides* (Siluriformes: Heptapteridae), a new catfish from the Río Ventuari, Venezuela. *Zootaxa* ,**2005,** 1009, 37–49.
Dorooshi, G. Catfish stings: A report of two cases. *J. Res. Med. Sci* . **2012, 17,** 578–581.
Egge, J.J.D.; Simons, A.M. Evolution of venom delivery structures in madtom catfishes (Siluriformes: Ictaluridae). *Biol. J. Linn. Soc.* **2011,** 102, 115–129.
Eiland ,L.S.; Salazar , M.L. Polymicrobial Catfish Spine Infection in a Child. *J. Pediatr. Infect. Dis*. 2006, 25 , 281–282
Ferlic ,R.J, ; Bonatz , E.; Robbin, M.L. Radial artery injury from a catfish sting. *Am. J. Orthop*. (Belle Mead NJ). **2003**, 32,412–414.
Fish, C.J.; Cobb, M.C. *Noxious Marine Animals of the Central and Western Pacific Ocean.* U.S. Government Printing Office, **1954.**
Fleming, J. *The Philosophy of Zoology: Or, A General View of the Structure, Functions, and Classification of Animals;* Nabu Press, **2010.**
Greven, H.; Flasbeck, T.; Passia, D. Axillary glands in the armoured catfish *Corydoras aeneus* (Callichthyidae, Siluriformes), *Verhandlungen der Gesellschaft für Ichthyologie,* **2006,** 65–69.
Haddad, V.; Martins, I.A. Frequency and gravity of human envenomations caused by marine catfish (suborder siluroidei): a clinical and epidemiological study. *Toxicon* **2006,** 47, 838–843.
Haddad, V.; Lima, C.; Lopes-Ferreira, M. Venomous Marine Fish: Osteichthyes (Bony Fish) . In: Gopalakrishnakone P. (eds) *Marine and Freshwater Toxin;* Springer, **2016.**
Haddad. V. *Medical Emergencies Caused by Aquatic Animals. A Zoological and Clinical Guide;* Springer Nature, **2020**..
Haddad, V.; de Souza, R.A.; Auerbach, P.S. Marine Catfish Sting Causing Fatal Heart Perforation in a Fisherman . *Wilderness Environ. Med.* **2008,** 19, 114–118.

Haddad, V. Injuries caused by fish in a community of Pantanal fishermen: detection, treatment, and prevention of envenomations and trauma. *Rev. Soc. Bras. Med. Trop.* .**2018,** 51,700–704.

Hardy , A.R.; Steinworth, B.M. ; Hale,M.E.. Touch sensation by pectoral fins of the catfish *Pimelodus pictus*. *Proc. Royal Soc.* B, **2016,** 283, 20152652.

Huang, G.; Goldstein, R.; Mildvan,D. Catfish spine envenomation and bacterial abscess with *Proteus* and *Morganella*: a case report . *J. Med. Case Rep*. **2013,** 7, 122

Junqueira, M.E.P.; Mondin, A.C.; Lopes, C.A.M. Microbiota characterization of the catfish (*Cathorops agassizii* and *Genidens genidens*) sting venom. *Braz. J. Vet. Res. Anim. Sci.* **2006,** 43, 793–796.

Kaar, C.R.J.; Plikaitis,C.; Germino, K.W.; Nakanishi, A.K. Catfish Noodling Forearm Injury Requiring Urgent Surgical Treatment: A Case Report and Review of the Literature. *Wilderness Environ. Med.* **2013**,7, 75–79.

Kapoor,B.G.; Khanna, B. *Ichthyology Handbook*. Springer Science & Business Media, **2004,** 1059 p.

Kumar, P.; Biswas , G.; Ghoshal , T.K.; Kailasam , M.; Christina , L.; Vijayan, K.K. Current knowledge on the biology, captive breeding and aquaculture of the brackish water catfish, *Mystus gulio* (Hamilton, 1822): A review. *Aquaculture* **2019,** 499, 243–250.

Macêdo, A.K.S.; Da Silva, J.R.P.; De Oliveira, S.P.; Haddad Jr.V. ; Vendel, A.L. Potentially dangerous fish of the Paraiba Estuary: Identification and envenomation mechanisms. *J. Coast. Life Med*. **2017, 5,** 459–462.

Manivasagan, P.; Annamalai, N.; Ashokkumar, S.; Sampathkumar, P. Studies on the proteinaceous gel secretion from the skin of the catfish, *Arius maculatus* (Thunberg, 1792). *Afr. J. Biotechnol.* **2009,** 8, 7125–7129.

Mann,J.W.; Werntz , J.R.J. Catfish stings to the hand. *Hand Surg Am.* **1991**, 16, 318–321.

McKinstry, D.M. Catfish stings in the United States: case report and review. *J. Wilderness Med.* **1993,** 4, 293–303.

Murphey, D.K.; Septimus, E.J.; Waagner, D.C. Catfish-Related Injury and Infection: Report of Two Cases and Review of the Literature. *Clin. Infect. Dis.* **1992**, 14, 689–693.

Negreiros, M.M.B.; Yamashita,S.; Sardenberg, T.; Fávero, Jr, E.L.; Ribeiro, F.A.H.; Haddad Jr.,W.T.; Haddad Jr., V. Diagnostic imaging of injuries caused by venomous and traumatogenic catfish. *Rev. Soc. Bras. Med. Trop.* **2016**, 49, 530–533.

Ng, H.H. *Akysis heterurus*, a new species of catfish (Teleostei: Akysidae) from Eastern Sumatra, with notes on *Akysis hendricksoni* Alfred. *Raffles Bull. Zool*. **1996**, 44, 3–10.

Ng, H.H. ; Rainboth, W.J. Four new species of *Akysis* (Teleostei : Siluriformes : Akysidae) from mainland Southeast Asia, with comments on *A. similis*. *Raffles Bull. Zool*. **2005**, 13, 33–42.

Ng, H.H. ; Kottelat, M. Description of three new species of catfishes (Teleostei: Akysidae and Sisoridae) from Laos and Vietnam. *J. South Asian Nat. Hist.* **2000,** 5, 7–15.

Pachaiyappan, A.; Sadhasivam, G.; Arumugam,S.; Muthuvel , A. Bio Prospecting the Enzymatic and Anticancer Potential of Spine Secretions of Marine Catfish (*Plotosus lineatus*). *J. Biol. Act. Prod. Nat.* **2015,** 5, 406–418,.

Pacy, H. Australian catfish injuries with report of a typical case. Med. J. Aust. 1966, 2, 63.

Pandey, S.; Upadhyay, R.K. Fish venom toxins: Natural source of pharmaceuticals and therapeutic agents "Pharmaceutical and therapeutic uses of fish venom toxins. *Int. J. Pharm. Pharm.* **2020,** 12, 1–14.

Park, J.; Kim,I. ; Kim, S. Structure and Histochemistry of the Skin of a Torrent Catfish, *Liobagrus mediadiposalis*. *Environ. Biol. Fishes* **2003,** 66, 3–8 .

Prithiviraj, N.; Annadurai, D. Studies On Bioactive Properties Of The Catfish *Plotosus canius* (Hamilton, 1822)Sting Venom And Epidermal Mucus. *Int. J. Recent Sci. Res.* **2010**,3, 467–473.

Ramos, A.D.; Conceição, K.; Silva Jr.PI. ; Richardson, M.; Lima, C.; Lopes-Ferreira, M. Specialization of the sting venom and skin mucus of *Cathorops spixii* reveals functional diversification of the toxins. *Toxicon* **2012**, 59, 651–665.

Reichert, L.; Worsham, J.; Fankhauser,G. Radial Artery Compression with Threatened Transection: Catfish Injury in a Pediatric Patient. *Ann. Vasc. Surg.* **2017,** 39, 288.e1-288.e3.

Sarmiento, B.E.; Rangel, M.; Gonçalves, J.C.; Pereira, L.; Rego, S.; Campos, L.A.; Haddad Jr. V.; Mortari, M.R.; Schwartz, E.F. First report of the characterization of the pathophysiological mechanisms caused by the freshwater catfish *Pimelodus maculatus* (order: Siluriformes). *Toxicon* 2015, 101, 55–62.

Satora , L.; Kuciel , M.; Gawlikowski, T. Catfish stings and the venom apparatus of the African catfish *Clarias gariepinus* (Burchell, 1822), and stinging catfish *Heteropneustes fossil*is (Bloch, 1794). *Ann. Agric. Environ. Med*, **2008**,15,163–166.

Satora , L.; Pach , J.; Targosz ,D.; Szkolnicka, B. Letter to the Editor: Stinging Catfish Poisoning. *Clin. Toxicol*. **2005,** 43, 893–894.

Sazima, I.; Zuanon, J.; Haddad, Jr.,V.. Puncture wounds by driftwood catfish during bucket baths: local habits of riverside people and fish natural history in the Amazon. *Wilderness Environ. Med.* **2005**, 16, 204–208.

Shiomi, K.; Takamiya ,M.; Yamanaka ,H.; Kikuchi, T.; Konno,K. Hemolytic, lethal, and edema-forming activities of the skin secretion from the oriental catfish (Plotosus lineatus). *Toxicon* **1986**, 24, 1015–1018.

Sodre,G.S; Colaço-Fernandes , T.R.; Rapp Py-Daniel, L.H.; Birindelli ,J.L.O.; Zuanon ,J. Putative poison gland in the thorny catfish Acanthodoras spinosissimus (Siluriformes: Doradidae) . *Acta Amazonica* **2021**, 51, 250–254.

Thalgaspitiya, S.P.B.; Wijerathne, B.T.B.; Siriwardhana,N.P. Freshwater Catfish Envenoming in a Tropical Country. *Wilderness Environ. Med.* **2017**, 28, 259–266 .

Usman, B.I.; Amin, S.M.N.; Arshad, A. ; Rahman, M.A. Review of Some Biological Aspects and Fisheries of Grey-Eel Catfish *Plotosus canius* (Hamilton, 1822). *Asian J. Anim. Vet. Adv*. **2013,** 8, 154–167.

Venkaiah ,Y. ; Lakshmipathi, V. Biochemical Composition of Epidermal Secretions and Poisonous Spine of Two Freshwater Catfishes. *Asian Fish. Sci.* **2000,** 13, 183–189.

Wellborn, T.L. *Channel Catfish - Life History and Biology*-. SRAC Publication No. 180, https://thefishsite.com/articles/channel-catfish-life-history-and-biology

White, J. ; Meier, J. *Handbook of Clinical Toxicology of Animal Venoms and Poisons;* CRC Press, **2017**.

Wright, J.J. *The evolutionary ecology of venomous catfishes, with a focus on members of the North American family Ictaluridae (Teleostei: Siluriformes)*; Ph.D. dissertation, University of Michigan, **2012**.

Wright, J.J. Diversity, phylogenetic distribution, and origins of venomous catfishes . *BMC Evol. Biol.* **2009,** 9, 282.

Wright, J.J.; Bailey, R.M. Systematic revision of the formerly monotypic genus *Tanganikallabes* (Siluriformes: Clariidae). *Zool. J. Linn. Soc.* **2012,** 165, 121–142.

Wright, J.J. Evolutionary History of Venom Glands in the Siluriformes, In : *Evolution of Venomous Animals and Their Toxins* (ed. Gopalakrishnakone, P. and Malhotra, A); Springer, **2015.**

Yoshino,T.; Kishimoto,H. *Plotosus japonicus*, a New Eeltail Catfish (Siluriformes: Plotosidae) from Japan. *Bull. Natl. Mus. Nat. Sci.,* Ser. A, Suppl. 2, **2008,** 1–11.

Zeman, M.G. Catfish stings: a report of three cases. *Ann. Emerg. Med.* **1989**,18, 211–213.

Ziegman, R.; Alewood , P. Bioactive Components in Fish Venoms. *Toxins* **2015**, 7, 1497–1531.

Links

Anon., https://whatdewhat.com/dangerous-freshwater-fish-can-kill/
Anon., http://extension.msstate.edu/content/biology-catfish
Anon., http://fishesofaustralia.net.au/home/species/2758
Anon., http://txstate.fishesoftexas.org/noturus%20gyrinus.html
Anon., http://www.fao.org/3/ad468e/ad468eJG.pdf
Anon., https://animaldiversity.org/accounts/Noturus_exilis/
Anon., https://animals.jrank.org/pages/2027/Catfishes-Siluriformes-CHANNEL-CATFISH-Ictalurus-punctatus-SPECIES-ACCOUNTS.html
Anon., https://en.wikipedia.org/wiki/African_bullhead
Anon., https://en.wikipedia.org/wiki/Gafftopsail_catfish
Anon., https://en.wikipedia.org/wiki/Pimelodus_pictus
Anon., https://en.wikipedia.org/wiki/Striped_Raphael_catfish
Anon., https://fishesofaustralia.net.au/home/species/2141
Anon., https://fishesofaustralia.net.au/home/species/2762
Anon., https://fishesofaustralia.net.au/home/species/3311
Anon., https://gcrl.usm.edu/public/fish/hardhead.catfish.php
Anon., https://whatdewhat.com/dangerous-freshwater-fish-can-kill/
Anon., https://www.aquariadise.com/raphael-catfish/
Anon., https://www.consultant360.com/article/catfish-spine-envenomation
Anon., https://www.consultant360.com/article/catfish-spine-envenomation
Anon., https://www.consultant360.com/article/catfish-spine-envenomation
Anon., https://www.consultant360.com/article/catfish-spine-envenomation
Anon., https://www.dovemed.com/diseases-conditions/catfish-sting/#
Anon., https://www.fishbase.de/summary/10220
Anon., https://www.fishbase.de/summary/288
Anon., https://www.fishbase.de/summary/ordersSummary.php?order=Siluriformes
Anon., https://www.fishbase.de/summary/plotosus-papuensis.html
Anon., https://www.fishbase.in/country/CountrySpeciesSummary.php?c_code=104&id=5140
Anon., https://www.fishbase.in/summary/12095
Anon., https://www.fishbase.in/summary/12099)
Anon., https://www.fishbase.in/summary/22590
Anon., https://www.fishbase.in/summary/4929
Anon., https://www.fishbase.in/summary/51770
Anon., https://www.fishbase.in/summary/958
Anon., https://www.fishbase.in/summary/Porochilus-argenteus.html
Anon., https://www.fishbase.se/summary/3032

Anon., https://www.fishbase.se/summary/3056
Anon., https://www.fishbase.se/summary/951
Anon., https://www.fishbase.se/summary/Bagre-marinus
Anon., https://www.fishbase.se/summary/Cathorops-agassizii.html
Anon., https://www.fishbase.se/summary/Clarias-fuscus
Anon., https://www.fishbase.se/summary/Neosilurus-ater
Anon., https://www.fishbase.se/summary/Netuma-bilineata.html
Anon., https://www.fishbase.se/summary/Noturus-fasciatus.html
Anon., https://www.fishbase.se/summary/Osteogeneiosus-militaris.html
Anon., https://www.fishbase.se/summary/Plotosus-canius
Anon., https://www.fishbase.se/summary/Plotosus-limbatus.html
Anon., https://www.fishbase.se/summary/Tandanus-tandanus
Anon., https://www.futurity.org/just-your-garden-variety-poisonous-catfish/
Anon., https://www.futurity.org/just-your-garden-variety-poisonous-catfish/
Anon., https://www.fws.gov/fisheries/ans/erss/uncertainrisk/ERSS_Heteropneustes_microps_final_January_2018.pdf
Anon.,https://www.gameandfishmag.com/editorial/identifying-fish-that-sting-and-maybe-kill-you/339189
Anon., https://www.leisurepro.com/blog/explore-the-blue/march-plotosus-lineatus/
Anon., https://www.liveabout.com/finned-by-a-saltwater-catfish-2929252
Anon., https://www.livescience.com/10615-1-000-catfish-species-venomous.html
Anon.,https://www.nps.org.au/australian-prescriber/articles/managing-injuries-by-venomous-sea-creatures-in-australia
Anon., https://www.sciencedaily.com/releases/2009/12/091210173607.htm
Anon., https://www.scotcat.com/articles/article119.html
Anon., https://www.scotcat.com/articles/article119.html
Anon., https://www.scotcat.com/articles/article84.htm
Anon., https://www.scotcat.com/factsheets/plotosus_lineatus.htm
Anon.,https://www.wbtv.com/2019/10/16/venomous-fish-with-painful-sting-was-presumed-extinct-smoky-mountains-its-back
Anon., U.S. Fish & Wildlife Service, 2017
Anon., https://www.sciencedaily.com/releases/2009/12/091210173607.htm
Rogers and Wojewoda, https://labmedicineblog.com/2018/05/15/microbiology-case-study-a-51-year-old-male-with-swelling-in-his-finger/
Wright .https://www.sciencedaily.com/releases/2009/12/091210173607.html
Wright, https://news.umich.edu/killer-catfish-venomous-species-surprisingly-common-study-finds/
Wright. http://www.nysm.nysed.gov/research-collections/biology/ichthyology/research/transcriptomic-identification-toxic-components
Wright,J.J. Transcriptomic Identification of Toxic Components of Catfish Venoms. http://www.nysm.nysed.gov/research-collections/biology/ichthyology/research/transcriptomic-identification-toxic-components.

INDEX

D

R

S

T

V

X